A Competitor's Compendium to the Geography Bee

I dedicate this book to my parents, brother, grandparents, relatives, friends, and teachers.

K.R.

A Competitor's Compendium to the Geography Bee

ISBN-10: 1976486513

ISBN-13: 978-1976486517

Printed in the United States

© Copyright 2015 Keshav Ramesh

© Copyright 2017, Third Edition

Font Set in Maiandra GD, Arabic Typesetting, Candara, Calibri, Ebrima

Summary: This is a compendium of geographic questions and vital information designed to prepare kids in grades 4-8 competing in the School, State, and National Geographic Bees, Junior and Senior NSF Geography Bees, United States Geography Olympiad, and International Geography Bee.

Design and Text by Keshav Ramesh

Cover Illustration by Keshav Ramesh

All rights reserved.

A Competitor's Compendium to the Geography Bee

A COMPETITOR'S COMPENDIUM TO THE GEOGRAPHY BEE

by Keshav Ramesh

A Competitor's Compendium to the Geography Bee

The Geography Bee Ultimate Preparation Guide and A Competitor's Compendium to the Geography Bee:

This young genius has come up with a great guide! His questions follow the various topics that come up on the bee and simulate the types of questions that come.

A must-buy for the Geography Bee contestant!

- Karan Menon, 2015 NGB Champion, 2015 U.S. Geography Olympiad JV Champion, and 2017 National U.S. International Geography Bee Third Place Winner

The Geography Bee Ultimate Preparation Guide:

These questions really helped me in my geography endeavors. Keshav is a great author who made many great questions. The content is very well organized and has many different tips and questions. The questions are easy to study from and can help you become smarter in no time!

- Rohit Gunda, 2016 Connecticut State Geographic Bee Third Place Winner

The Geography Bee Ultimate Preparation Guide and A Competitor's Compendium to the Geography Bee:

Your books are so useful! I got 3rd at the NJ state bee using just them and travel videos and an atlas. I encourage you to check out books written by Keshav Ramesh, which each has specific sections designed for different levels of the bee. I really appreciated Keshav's book, and it taught me a lot of things. It's a very good book.

-Ken Mitchell, 2017 New Jersey State Geographic Bee Third Place Winner

A Competitor's Compendium to the Geography Bee

Table of Contents

Tips, Tricks, and How to Prepare for the Geography Bee 8

North America 26

    United States 26

    Canada, Greenland, and Mexico 63

    Central America 86

    The Caribbean 94

South America 101

    Northwestern South America 101

    Southern South America 108

    Brazil and the Guianas 114

Asia 120

    South Asia 120

    Southeast Asia 149

    East Asia 167

    Central Asia and Asian Russia 180

    Middle East 188

    Turkey and the Caucasus 202

Europe 217

    Northern Europe 217

Western Europe	223
Southern Europe	234
Central Europe	242
Eastern Europe and European Russia	245
Africa	**251**
Northeast Africa	251
Southern Africa	259
Central Africa	269
North Africa	276
West Africa	281
Australia and Oceania	**290**
Australia	290
Melanesia and Papua New Guinea	295
Polynesia and New Zealand	300
Micronesia	304
Antarctica	**307**
Mock Bee	**310**
Classroom/School Competition:	310
State Qualification Test	314
State Competition	319
Round 1: The United States	319
Round 2: Cultural Geography	321
Round 3: Physical Geography	323

A Competitor's Compendium to the Geography Bee

    Round 4: Economic Geography 325

    Round 5: The World 328

    Round 6: Political Geography 330

    Round 7: U.S. National Parks 332

    Round 8: Major Cities 334

   National Competition Preliminaries 337

2013 National Geographic Bee Finals Competition 341

2015 National Geographic Bee Finals Competition 356

USA Geography Olympiad/iGeo Resources 366

Geo Statistics 369

About the Author 379

Bibliography 380

A Competitor's Compendium to the Geography Bee

Tips, Tricks, and How to Prepare for the Geography Bee

What materials should I have when studying geography?

When studying for the geography bee, I would recommend keeping an atlas, detailed political map, and detailed physical map near you. These three are the most vital sources you need to help you achieve success in the National Geographic Bee.

From looking at a detailed political map, you can find cities and countries from around the world. Territories, islands, and dependencies will be included. You can find oceans, seas, lakes, straits, gulfs, bays, and even rivers. These will be emphasized in a physical map.

There are also smaller versions of political maps – for example, ones showing just South Asia or Central America.

Archipelagos are scattered across the world, like Indonesia and Japan. A political map will show you the countries in different colors, and they will be labeled.

Political maps will also show you the continents, and the names of countries will be printed much bigger than the cities.

A physical map will also help as well, to identify the landforms (which will be labeled) as well as biomes. Ocean ridges and seamounts can be displayed on physical maps but not necessarily on all.

Different colors will be expressed on the map, as well as ridges to show mountains, or a plain, yellowish color spreading across a certain area to signify a desert.

Also shown on physical maps are bodies of water, plateaus, geographical regions, basins, and other major parts of the Earth's topography.

In detailed physical maps, or ones that show a close-up of a region or country in the world, also have reservoirs, smaller bays/gulfs, highlands, hills, passes, and much more.

Atlases are great for researching thousands of facts just by looking at maps and diagrams, and reading about them.

A Competitor's Compendium to the Geography Bee

Countries are always featured in atlases, and you'll always find a few chapters about the physical geography of the world.

You will find many thematic maps in atlases, such as those depicting religions, language families, economy, water supply, major agricultural commodities, and more.

I would recommend buying atlases sold by National Geographic, as they, in my opinion, provide the best information.

The National Geographic Atlas of the World, Tenth Edition is good for participants serious about winning the state bees and doing well in the national bees. There is no atlas like this in the world that will have a lot of very detailed information. Although it is expensive, it is worth the price. State winners have been awarded this book in the past.

You should get an almanac, like the *National Geographic Kids Almanac 2018* or another good one recently published from National Geographic with interesting (and sometimes weird!) geography facts.

A Competitor's Compendium to the Geography Bee

However, this book gives you a broader outlook on geography, so it is a good reference for your school and sometimes even your state bees.

The *National Geographic Magazine* is especially important, as questions regarding facts in the magazine have been asked.

If you are participating in the 2018 National Geographic Bee, learn the geographical facts from the magazine from its six issues before May (November 2017-April 2018). This applies to 2019 and beyond as well.

You should also get a book about the physical geography of the world, as this is vital to the physical geography section of the National Geographic Bee. The *National Geographic Desk Reference* is probably the best option for you when looking at physical geography.

How difficult are the questions in this book?

The difficulty of these questions range from the State Qualification Test to the final rounds of the National Bee. You should use these questions to prepare to do well in and/or win your state and national bee.

A Competitor's Compendium to the Geography Bee

The questions increase in difficulty as you finish more and more, before matching questions found at the final levels of the state bee and the preliminary rounds of the national bee.

Sometimes the questions will be easier than the preceding ones in this book, but it is good to be prepared for whatever question is directed at you during any stage of the competition.

In other words, only some of the questions in this book will be in order of difficulty, and most of the questions will be in no specific order.

How many questions are in this book?

A Competitor's Compendium to the Geography Bee has **over 2,025 questions to help you in all levels of the National Geographic Bee!**

This book focuses on the specific regions on all of the world's continents, and the questions are crafted to help you win any stage of the competition!

What is the daily amount of time I should study?

A Competitor's Compendium to the Geography Bee

If you want to win the school bee, studying for 30 minutes to 1 hour a day is enough (unless you are in a competitive school for the geography bee for those of you in New Jersey, Florida, California, Washington, Michigan, Virginia, Maryland, or Texas). For the state level bee, I would recommend studying 2 to 3 hours a day.

If you are at the level of the National Bee Preliminaries, 3 to 4 hours is what I would recommend. Those aspiring to achieve a place in the top ten should definitely amount to more than 4 hours, if not 5-7.

This competition, like others, is challenging. Hours of dedication to geography is vital to your success in the state and national bees.

Is this book only for competitors in the National Geographic Bee?

This book is not only for people competing in the National Geographic Bee, but for other competitions as well.

If you are participating in the finals of the **North South Foundation's Junior/Senior Geography Bees**, this is a good

A Competitor's Compendium to the Geography Bee

book for you. There is a separate chapter for Indian Geography.

This book contains questions not only written for the National Geographic Bee, but also for the **United States Geography Olympiad (USGO)**. The USGO starts regionally as a National Qualifying Exam and any competitor must score above the national median score and/or in the top half at the regional level.

The **National USGO Championships** are held along with the **National History Bee and Bowl,** and the U.S. International Geography Bee in Arlington, Virginia annually in April. This Olympiad selects the top four individuals in the Varsity Division to represent the United States at the **International Geography Olympiad (iGeo)**, held in a different city around the world annually.

This book has also been written for students preparing for the **International Geography Bee (IGB)**, an intense geography competition created in 2017 with a regional National Qualifying Exam and fast quiz bowl rounds at the national level.

The U.S. national championships of the International Geography Bee is held in conjunction with the USGO and the

A Competitor's Compendium to the Geography Bee

National History Bee and Bowl annually in Arlington, Virginia in April. The top four individuals in the JV Division represent the United States region at the **Junior Varsity IGB World Championships.** The top four individuals in the Varsity division represent the United States region at the **Varsity IGB World Championships.**

As a list, this book is a great resource for these competitions:

- National Geographic Bee (NGB)
- North South Foundation Junior/Senior GeoBees
- United States Geography Olympiad (USGO)
- International Geography Bee (IGB)

What should I study for the state and national competitions of the National Geographic Bee?

For the state bee, knowledge of borders, locations, and the United States is important. Be sure to know your national parks, national forests, and national monuments. You should be well versed in the major cities of each state and country.

Be sure to follow National Geographic's Instagram and Twitter for more geographic information. **In the 2016 State Geographic Bees, questions regarding photos from National**

A Competitor's Compendium to the Geography Bee

Geographic's Instagram account were posed to the competitors in the final rounds.

If you reach the finals of the state bee, your atlas should be your greatest ally. You should have a complete mastery or near mastery of thousands and thousands of locations – however, this should not be achieved through memorization, but through constant review. It may sound difficult, but it gets much easier after you practice and practice, at least three hours a day.

Although knowing thousands of locations sounds like a lot, you probably know the name and location of every single country, some major cities, major rivers, major lakes, major mountain ranges, notable islands, deserts, oceans, and seas if you are an experienced geography bee participant. This by itself is a lot of information.

Be thorough with your atlases, and use the National Geographic Atlas of the World, Tenth Edition as much as possible.

Create questions, research facts, connect locations, and take notes. Keep rereading what you have read and written down/typed and try to produce questions from the facts.

Integrate the facts and questions as much as possible so that you can remember them.

You should be thorough with tourist attractions and landmarks, as well as river confluences, port cities, UNESCO World Heritage Sites, major exports, current events, physical geography terms, and geographic extremes (such as Kanyakumari, Cape Byron, and Ushuaia).

What do I need to know about each country?

Here's a list of what you need to know about each country to help you prepare for the National Geographic Bee. Remember, some of the things on this list may not apply to certain countries:

Basics:
- Location (Continent)
- Location (Region, e.g. South Asia)
- Capital
- Major Cities (At least 10+ for countries with populations over 70,000,000)
- Population (Approximate)
- Official Country Name

Physical:

A Competitor's Compendium to the Geography Bee

- Highest and Lowest Points
- Mountain Ranges, Peaks, and Volcanoes
- Rivers, Deltas, River Mouths/Sources
- Lakes, Reservoirs, and Dams
- Bordering Seas
- Gulfs and Bays
- Straits, Sounds and Inlets
- Plateaus
- Peninsulas, Capes, and Points
- Plains and Basins
- Wetlands, Swamps, and Marshes
- Deserts, Ergs, and Dunes
- Valleys and River Valleys
- Grasslands and Prairies
- Waterfalls
- Canyons and Gorges
- Islands and Archipelagoes
- Isthmuses and Spits
- Canals
- Physical Regions
- Buttes and Mesas
- Major Glaciers and Fjords
- Lagoons and Reefs
- Continental Divides, Earth Physical Structure, Layers of the Earth

Political

A Competitor's Compendium to the Geography Bee

- Bordering Countries
- Administrative Divisions (States, Provinces, Federal Districts, Counties)
- Territories, Dependencies, and Occupied Atolls
- Current Leader(s) (President, Prime Minister, King/Queen, Chairman, Prince, etc.)
- Government Structure/Important and Influential Laws and Type of Rule (Republic, Democracy, Monarchy, etc.)
- Disputed Countries and Regions
- Politically Established Regions (Northeast Africa, Central America, etc.)
- National and Global Organizations

Cultural
- Religions
- Languages
- Festivals, Holidays, and Traditions
- Foods, Art, Music, and Architecture
- Cultural Items/Objects and Symbols

Environmental
- Oceans
- Conservation and Biodiversity
- Biomes and Habitats
- Plants and Animals
- Global Warming/Climate Change

A Competitor's Compendium to the Geography Bee

- Environmental Hot Spots
- Natural Disasters

Economic
- Currencies
- Trade, Exports, and Imports
- Production
- Agricultural Products and Natural Resources
- Port Cities, Seaports, and International Airports

Historical:
- Kingdoms and Empires
- Wars
- Former Countries, Historical Colonies, Territories, Past Leaders, Independence

Landmarks
- National Parks/Preserves, Forests, and Monuments
- National Historic Sites and Historical Parks (United States)
- UNESCO World Heritage Sites
- Famous Castles/Ruins, Museums, and Zoos
- Space Centers, Observatories, and National Laboratories

Current:
- Global, Nationwide, and Environmental Issues

A Competitor's Compendium to the Geography Bee

- Archaeological Discoveries
- Climate Change and Global Warming
- International Relations
- Major Worldwide Sporting Events (FIFA World Cup Russia 2018, Tokyo 2020 Summer Olympics)

What are some good websites I can use to prepare?

The website I'd recommend to help you prepare for the competition is www.nationalgeographic.com/geobee. It has great tools to help you study, and also a page where you can play a game called the GeoBee Challenge, where National Geographic gives you 10 new questions every day to prepare with.

Sites Dedicated Exclusively to Geography and the Geography Bee:

www.nationalgeographic.org/geobee

www.geobeeworld.blogspot.com

www.geography.about.com

http://lizardpoint.com/geography/

http://www.sporcle.com/games/category/geography/all

A Competitor's Compendium to the Geography Bee

http://nationalgeographic.org/bee/study/play-kahoot/

What are some other online resources I can use?

Quizlet is an online quizzing program that you can use to prepare for the geography bee. I would also use Socrative, where you can make multiple choice questions with multiple answers (if you choose). There are hundreds, maybe thousands of geography quizzes on Sporcle.

The world's largest geography bee community is on Google+. Look up **GeoBee City** on Google+, and you will find it. GBC is a geography bee community where students from all over the country work on geography-related projects, mock bees, quizzes, maps, and use resources to study. GBC was founded by Karan Menon, the 2015 National Geographic Bee Champion, in 2014 and the community has over 200 members.

In addition to studying for the National Geographic Bee with fellow students, we prepare for the **United States Geography Olympiad** and the **International Geography Bee**, and some of us who are eligible prepare for the **North South Foundation (NSF) Junior and Senior Geography Bees.**

A Competitor's Compendium to the Geography Bee

Any questions or concerns?

Any questions you'd like me to add? Still need help with geography? Any errors that need to be fixed? Contact me at keshav.ramesh@gmail.com!

Is there anything else I should know when preparing for the geography bee?

There is!

Let us set the scenario that you were participating in the 2012 State Geographic Bee in fourth grade, up to the 2016 State Geographic Bee in eighth grade. This way, National Geographic has a near-perfect way of guaranteeing that you wouldn't have known some of the questions asked in the 2011 State Geographic Bee finals as you would've been in third grade and most likely not watched the final rounds.

Let us look at a different scenario. Let's say you were participating in the 2013 State Geographic Bee in fourth grade, up to the 2017 State Geographic Bee in eighth grade. Again, wouldn't National Geographic have a near-perfect way of guaranteeing that you wouldn't have known some of the questions asked in the 2012 State Geographic Bee? So instead of creating a set of new final round questions, they could just reuse them.

Certainly, the National Geographic Society would have to create new questions for the state finals – but not all of the questions would necessarily have to be created as of that year.

So for those of you participating in the 2018, 2019, or 2020 State Geographic Bee, I would suggest looking at these videos below now or sometime later, and make sure that you record the questions, test yourself on them, and maybe find some other patterns between the questions of different years:

2013 for 2018:

http://ct-n.com/ctnplayer.asp?odID=8880

2014 for 2019:

https://www.youtube.com/watch?v=rJX_YSSOVU8

2015 for 2020:

https://www.youtube.com/watch?v=Tu3xJLqJIFA

2016 for 2021:

https://www.youtube.com/watch?v=6FuT9nAl9dY

2017 for 2022:

http://ct-n.com/ctnplayer.asp?odID=13916

Should I use National Geographic's Instagram account to help me study?

Using National Geographic's Instagram account to learn geography is a great idea. (you don't need an Instagram account to do this!).

Many questions are actually based off of pictures National Geographic posts of oceans, animal and plant species, conservation projects, geography, science, and nature. Check out these links!:

https://www.instagram.com/natgeo/

https://www.instagram.com/natgeotravel/

https://www.instagram.com/natgeoyourshot/

https://www.instagram.com/natgeowild/

Good luck and happy studying for the National Geographic Bee!

A Competitor's Compendium to the Geography Bee

A Competitor's Compendium to the Geography Bee

North America

United States

1. Havre and Great Falls are cities in what U.S. State bordering North Dakota and Canada?
Montana

2. The Near Islands belong to what U.S. State?
Alaska

3. Cupertino can be found in what U.S. State that is home to the Mojave Desert?
California

4. The majority of French-speaking people in the United States live in what state bordering Arkansas to the north?
Louisiana

5. Pocatello is a city near the Snake River in what U.S. State?
Idaho

6. Sault Ste. Marie is at the tip of what peninsula bordering Lake Superior and Lake Huron?
Upper Peninsula

7. Galveston is a city located miles southeast of what major Texan city?

Houston

8. Lake Okeechobee and West Palm Beach are located in what U.S. State?
Florida

9. Coyotes originated in what geographical region of the United States?
Southwest

10. Hawaii, although part of the United States, is geographically part of what region in Oceania?
Polynesia

11. What U.S. State is known as the lightning capital of the United States?
Florida

12. Chrysler Building is located in Manhattan in what U.S. State?
New York

13. Castle Geyser is a famous geographical attraction part of what national park?
Yellowstone National Park

14. Hyperion, the world's tallest living tree, is taller than the Statue of Liberty. It is located in Redwood National Park in what U.S. State known as "The Golden State"?
California

15. The Aleutian Mountains are located on what peninsula in Alaska?
Aleutian Peninsula

16. Mount St. Helens is a famous peak in what mountain range extending from northern California to Washington?
Cascade Mountains

17. The Colorado Plateau is located between the Cascade Mountains and what basin?
The Great Basin

18. The Everglades is a swampy region located on what peninsula in the southeastern United States?
Florida

19. The Chilkoot Pass crosses from the United States into what country?
Canada

20. The Strait of Juan de Fuca separates the United States from what country?
Canada

21. The Gulf of Alaska feeds into what ocean?
Pacific Ocean

22. Lake Pontchartrain, in the southern United States, is the largest lake in what U.S. State?
Louisiana

23. The Rio Grande, a river forming much of the border between the United States and Mexico, has its source in what major mountain range?
Rocky Mountains

24. The Colorado River, which forms the Grand Canyon, empties out into what gulf with the same name as a U.S. State?
Gulf of California

25. Yosemite Falls is located in what major mountain range in California?
Sierra Nevada

26. Glen Canyon is a hydroelectric dam on the Colorado River creating what major lake?
Lake Powell

27. The Seward Peninsula borders the Bering Strait, Bering Sea, and what other sea?
Chukchi Sea

28. The Mississippi River Delta is located in what U.S. State with the cities of New Orleans and Baton Rouge?
Louisiana

29. The Snake River forms Hells Canyon, the deepest gorge in the United States and is a tributary of what river forming much of Oregon's border with Washington?
Columbia River

30. Cape Hatteras is a chain of barrier islands in the Atlantic Ocean off the coast of what U.S. State?
North Carolina

31. Kingman Reef is a territory of the United States in what ocean?
Pacific Ocean

32. Houston is a major port city on what body of water east of Mexico?
Gulf of Mexico

33. Los Angeles is a city on the edge of the Coast Ranges in what U.S. State?
California

34. Cape Flattery can be found at the northwestern tip of what peninsula in Washington?
Olympic Peninsula

35. Phoenix is a city in Arizona on the edge of what major desert?
Sonoran Desert

36. Katmai National Park is located in what U.S. State where you can view the Aurora Borealis?
Alaska

37. The Susquehanna and Allegheny Rivers are located in what U.S. State where the cities of Allentown and Lancaster can be found?
Pennsylvania

38. Yale University is an Ivy League in what U.S. State bordering Long Island Sound?
Connecticut

39. The Catskill and Adirondack Mountains are located in what U.S. State?
New York

40. Mt. Frissell is the highest point in what U.S. State bordering New York and Massachusetts?
Connecticut

41. Lake Champlain is located in New York and what U.S. State whose capital is Montpelier?
Vermont

42. The Finger Lakes and the Mohawk River are in what U.S. State whose capital is Albany?
New York

43. The easternmost point in the United States is located in what U.S. State bordering Quebec and New Brunswick?
Maine

44. Lake Chocurua is located in the White Mountains in what U.S. State?
New Hampshire

45. Mystic Seaport is located in what U.S. State home to the Housatonic and Naugatuck Rivers?
Connecticut

46. Prime Hook National Wildlife Refuge is located in what U.S. State bordering Delaware Bay?
Delaware

47. Acadia National Park is located in what U.S. State including Mooselookmeguntic Lake and Sebago Lake?
Maine

48. The Chesapeake Bay Bridge is located in what U.S. State whose capital is Annapolis?
Maryland

49. What language is the second most spoken language in the United States?
Spanish

50. The Salton Sea is north of what valley in southern California, west of the Colorado River?
Imperial Valley

51. Mt. Elbert is located in what major mountain range west of the Park and Front Ranges?
Rocky Mountains

52. The Davis Mountains are north of the Rio Grande in what U.S. State?
Texas

53. The Pearl River is located in what U.S. State bordering Alabama?
Mississippi

54. The Yellowstone River, Milk River, and Flathead Lake are located in what U.S. State?
Montana

55. Lake of the Ozarks is north of the Ozark Plateau in what U.S. State straddling part of the Missouri River?
Missouri

56. Isle Royale is an island belonging to what U.S. State bordering Ontario?
Michigan

57. The Des Moines River flows through what U.S. State that is part of the Central Lowland and borders Minnesota and Nebraska?
Iowa

58. The Allegheny Plateau is west of what mountain range through which the Susquehanna River flows?
Appalachian Mountains

59. Long Island Sound is south of what U.S. State bordering Rhode Island?
Connecticut

60. The Toledo Bend Reservoir is on Texas's border with what U.S. State bordering Atchafalaya Bay?
Louisiana

61. The Cumberland Plateau is east of the Cumberland and Tennessee Rivers. This plateau is also west of what major mountain range?
Appalachian Mountains

62. Point Barrow is a city in what U.S. State where the Tanana and Kuskokwim Rivers can be found?
Alaska

63. Lanai is an island in what U.S. State home to the peak of Mauna Kea?
Hawaii

64. The Colorado Plateau encompasses Colorado, Utah, Arizona, and what U.S. State?
New Mexico

65. The Edwards Plateau is located north of the Rio Bravo del Norte in what U.S. State?
Texas

66. Georgian Bay is an inlet of which of the five Great Lakes?
Lake Huron

67. Apalachee Bay is south of what U.S. State straddling part of the Gulf Coastal Plain?
Florida

68. The St. Johns River is located in what U.S. State where Cape Sable can be found?
Florida

69. The Sangre de Cristo Mountains are located in Colorado and what U.S. State home to the Elephant Butte Reservoir?
New Mexico

70. The Sacramento Mountains are located in what U.S. State whose capital is Santa Fe?
New Mexico

71. Moosehead Lake is located in what U.S. State bordering the Bay of Fundy?
Maine

72. The Green Mountains are to Vermont as the White Mountains are to what?
New Hampshire

73. What river, named after a U.S. State in this region, is the longest river in New England?
Connecticut River

74. The Connecticut River forms the border between what two U.S. States?
Vermont and New Hampshire

75. The Flint Hills are located in what U.S. State home to the Smoky Hill River?
Kansas

76. The Yazoo River is located in what U.S. State bordering the Gulf of Mexico and Tennessee?
Mississippi

77. The Sacramento Valley is located in what U.S. State bordering Monterey Bay?
California

78. The Channel Islands and the Sierra Nevada are in what state?
California

79. Harney Peak is in the Black Hills in what state west of Minnesota?
South Dakota

80. The Ruby Mountains are located in what state home to the Humboldt River?
Nevada

81. The Olympic Mountains are located in what state?
Washington

82. The Connecticut River empties out into what sound?
Long Island Sound

83. The Nushagak Peninsula is located north of Bristol Bay in what state?
Alaska

84. Craters of the Moon National Monument is located in what state?
Idaho

85. Albemarle Sound and Pamlico Sound are on what coast of the United States?

East Coast

86. The Cortez Mountains and Ruby Mountains can be found in what U.S. State bordering Oregon and Idaho?
Nevada

87. Golden Spike National Historic Site is located northeast of the Great Salt Lake in what state?
Utah

88. Arabic, Chinese, English, French, Russian, and what other language are the official language of the United Nations?
Spanish

89. Ban Ki-moon is the current secretary-general of what major world organization?
United Nations

90. The Sonoran Desert is in what state whose highest natural point is Humphreys Peak?
Arizona

91. Lake St. Clair borders what state?
Michigan

92. The Keweenaw Peninsula belongs to what state bordering Lake Michigan?
Michigan

93. The most populous metropolitan area in the United States is at the mouth of the Hudson River. Name this city.
New York City

94. Wilkes-Barre is a city in what state bordering New York?
Pennsylvania

95. The Boston Mountains are south of what plateau?
Ozark Plateau

96. Haleakala National Park is located on what Hawaiian Island?
Maui

97. The Trinity Islands belong to what state bordering Canada to the east?
Alaska

98. The Niagara River cascades over what waterfall also known as Canadian Falls and part of Niagara Falls?
Horseshoe Falls

99. The Shumagin Islands belong to what state?
Alaska

100. Murfreesboro is a major city in what state?
Tennessee

101. Lafayette is one of the largest cities in what state bordering Timbalier Bay?
Louisiana

102. The Glass Mountains and Chisos Mountains are located in what state?
Texas

103. Athens is a major city in what state bordering the Atlantic Ocean and South Carolina?
Georgia

104. Lake Marian is located in what state whose capital is Columbia?
South Carolina

105. Casper is a major city in what state bordering Montana and Idaho?
Wyoming

106. The Caballo Reservoir and Brantley Lake are located in what state whose most populous city is Albuquerque?
New Mexico

107. The Endicott Mountains are west of the Philip Smith Mountains in what state bordering Camden Bay and Yakutat Bay?
Alaska

108. Great Quittacas Pond is located in what state bordering Nantucket Sound?
Massachusetts

109. Waterbury and New Haven are major cities in what state bordering the Atlantic Ocean to the south?
Connecticut

110. Lake Winnipesaukee is located in what state bordering Vermont and Massachusetts?
New Hampshire

111. Pictured Rocks National Lakeshore Isle Royale National Park are located in what state bordering Grand Traverse Bay?
Michigan

112. The Great Lakes Naval Training Center is located in what state bordering Lake Michigan to the northeast?
Illinois

113. Mammoth Cave National Park is located in what state south of Ohio?
Kentucky

114. Lake Wappapello is located in what state bordering Iowa, Illinois, and Kansas?
Missouri

115. Aransas National Wildlife Refuge is located in what state with the major cities of San Antonio and El Paso?
Texas

116. Mississippi borders what sound to the south?
Mississippi Sound

117. Sabine National Wildlife Refuge is located in the southwestern region of what state bordering Terrebonne Bay?
Louisiana

118. The Ross Barnett Reservoir is located in what state bordering Arkansas to the northwest and Louisiana to the southwest?
Mississippi

119. Shenandoah National Park is located in what state bordering Chesapeake Bay?
Virginia

120. Cuyahoga Valley National Park is located in what state?
Ohio

121. Assateague Island National Seashore is shared by Virginia and what state?
Maryland

122. Albemarle Sound is located in what state bordering Tennessee and Virginia?
North Carolina

123. Paterson and Newark are major cities in what state bordering Delaware Bay to the south?
New Jersey

124. Saginaw Bay, which borders Michigan, is located in what lake?
Lake Huron

125. Canandaigua Lake is located in what state bordering the Canadian province of Ontario to the west?
New York

126. Penobscot Bay borders what state with the cities of Auburn and Kennebunk?
Maine

127. Noatak National Preserve and the Schwatka Mountains are located in what state bordering the Hinchinbrook Entrance and Blying Sound?
Alaska

128. The Kohala Mountains are located in what island in Hawaii that is home to the Onizuka Center for International Astronomy?
Hawaii

129. The Pailolo Channel separates Maui from what island where the Kauhako Crater can be found?
Moloka'i

130. The Salinas Pueblo Missions National Monument is located in what state bordering Arizona and Colorado?
New Mexico

131. Fort Bliss is located miles away from El Paso in what state?
Texas

132. Petrified Forest National Park is located in what state whose capital is Phoenix?
Arizona

133. Black Canyon of the Gunnison National Park is located in what state bordering Nebraska?
Colorado

134. The Uinta Mountains and city of Provo are located in what state?
Utah

135. The Salton Sea is located southwest of the Orocopia Mountains in what state bordering the Gulf of Santa Catalina?
California

136. The Antelope Valley and the Tehachapi Mountains are located in what state where the Walker Pass can be found?
California

137. Goose Lake is shared by California and what other state with the major cities of Salem and Portland?
Oregon

138. The Clan Alpine Mountains are located in the western region of what state whose major cities include Las Vegas and Carson City?
Nevada

139. Point Arguello and Point Conception are located in what state bordering the Pacific Ocean?
California

140. The Medicine Bow Mountains are located in Wyoming and what other state with the cities of Westminster and Denver?
Colorado

141. The Wind River Range is located northwest of the Great Divide Basin in what state?
Wyoming

142. Timpanogos Cave National Monument is located north of Utah Lake in what state?
Utah

143. The Klamath Mountains are located in Oregon and what other state?
California

144. The Patuxent River Naval Air Test Center is located in what state?
Maryland

145. Lake Champlain is shared by New York, Quebec, and what state?
Vermont

146. The Potomac River empties out into what major bay in the Atlantic Ocean?
Chesapeake Bay

147. Lubbock and Beaumont are cities in what state?
Texas

148. Lake Winnebago is located in what state bordering Lake Michigan?
Wisconsin

149. The Altamaha River empties out into what ocean?

Atlantic Ocean

150. The Harry S. Truman Reservoir is located in what landlocked state bordering Iowa to the north?
Missouri

151. Harney Peak is located in the Black Hills in what state?
South Dakota

152. Eufaula Lake is located in what state known for its many tornadoes?
Oklahoma

153. Yellowstone Lake is located in what state with the cities of Laramie and Casper?
Wyoming

154. The Humboldt River is located in the northern region of what state partially in the Great Basin?
Nevada

155. Cape Blanco juts out into the Pacific Ocean in what state, forming its westernmost point?
Oregon

156. Misty Fjords National Monument is located in what state bordering the Cook Inlet?
Alaska

157. Malaspina Glacier, the largest piedmont glacier in the world, is located in what state?
Alaska

158. Lake Winnibigoshish is located north of Leech Lake in what state?
Minnesota

159. Table Rock Lake is located in the Ozark Plateau in what state?
Missouri

160. Cape San Blas is located in what state bordering Tampa Bay?
Florida

161. Magazine Mountain is located in what state?
Arkansas

162. Cape Lookout and Cape Fear are located in what state?
North Carolina

163. Spruce Nob is a peak in the Allegheny Mountains in what state?
West Virginia

164. Mount Rogers and Black Mountain are located in the eastern part of what state?
Kentucky

165. Mount Marcy is located in the Adirondack Mountains in what state?
New York

166. Behoboth Bay is located in what state?

Delaware

167. The Delmarva Peninsula can be found in what state whose capital is Annapolis?
Maryland

168. The Delaware River forms the entire border between New Jersey and what other state?
Pennsylvania

169. The Susquehanna River empties out into what major bay?
Chesapeake Bay

170. Oconee National Forest and Chattahoochee National Forest are located in what state?
Georgia

171. The Fort Peck Indian Reservation is located in what state bordering Saskatchewan and Alberta?
Montana

172. Bombay Hook National Wildlife Refuge is located in what state?
Delaware

173. Stratford Point, which extends out into Long Island Sound, is located in what state?
Connecticut

174. The Chandeleur Islands and Marsh Island belong to what state?
Louisiana

175. The Los Alamos National Laboratory is located in what landlocked state bordering Colorado?
New Mexico

176. The Crazy Horse Memorial, honoring the culture and heritage of the Native Americans and a Lakota chief, is located in what state?
South Dakota

177. Most Mormons in the United States live in what state whose capital is Salt Lake City?
Utah

178. The Desert National Wildlife Range is located in what state bordering Arizona?
Nevada

179. The Grand Coulee Dam is located in what northwestern state?
Washington

180. The Gateway Arch is located in St. Louis in what state?
Missouri

181. The Jefferson Davis Monument is a famous site in what state?
Kentucky

182. The San Jacinto Monument is located in what state where you can find many Spanish speakers?
Texas

183. Perry's Victory and International Peace Memorial is located in what state?
Ohio

184. Apalachicola National Forest is located south of Tallahassee in what state?
Florida

185. The Ozark Plateau is located in Missouri and covers much of what other state?
Arkansas

186. What two languages are spoken the most in the United States?
English and Spanish

187. What religion is the second largest in the United States?
Judaism

188. The Mississippi River has its source in what lake in Minnesota?
Lake Itasca

189. The Saint Elias Mountains, in the United States, extend into what country?
Canada

190. The Columbia Plateau is located between the Cascade Range and what major mountain range?
Rocky Mountains

191. The Everglades cover much of the southern part of what major peninsula in the south?
Florida Peninsula

192. The Sonoran Desert is located in what U.S. state?
Arizona

193. The Mississippi River Delta, which feeds into the Gulf of Mexico, is located in what state?
Louisiana

194. Lake Okeechobee is the largest lake in what state?
Florida

195. The Little River Canyon National Preserve is located in what state bordering Mississippi?
Alabama

196. Mount Davis is the highest point in what state bordering New York and Ohio?
Pennsylvania

197. The Germany Valley is located in what state bordering Virginia?
West Virginia

198. The Motorsports Hall of Fame, adjacent to the Talledega Superspeedway, is located in what state?
Alabama

199. What state ranks first among the 31 states that produce nuclear power in the United States?

A Competitor's Compendium to the Geography Bee

Illinois

200. What city in Connecticut is known as the insurance capital of the United States?
Hartford

201. The Roger Williams National Memorial is located in what state?
Rhode Island

202. Lake Marion and Lake Moultrie are located in what state bordering Long Bay?
South Carolina

203. Afognak Island is located in what state bordering Norton Sound?
Alaska

204. Ozark National Forest is located south of Norfork Lake in what state?
Arkansas

205. The Delaware Water Gap National Recreation Area is east of Stroudsburg, a city in what state?
Pennsylvania

206. Monongahela National Forest is located in what landlocked state bordering Kentucky and Ohio?
West Virginia

207. Willapa Bay and Grays Harbor border what state whose largest city is Seattle?

Washington

208. The Vermilion National Wild and Scenic River is located in what state bordering Iowa and Lake Michigan?
Illinois

209. The Cimarron National Grassland is located in what state whose largest city is Wichita?
Kansas

210. Chippewa National Forest is located in what state bordering Ontario and Manitoba?
Minnesota

211. Aurora and Denver are major cities in what state?
Colorado

212. The Osage Nation Reservation and the Black Kettle National Grassland are both located in what state?
Oklahoma

213. Redwood National Park and Joshua Tree National Park are located in what state bordering Monterey Bay?
California

214. The Ohoco Mountains are located in what state whose most populous city is Portland and whose capital is Salem?
Oregon

215. Chequamegon-Nicolet National Forest is located in what state with the Door Peninsula?
Wisconsin

216. Grand Teton National Park is located in what state bordering Colorado and Nebraska?
Wyoming

217. The Juniata River is located in what state whose capital is Harrisburg?
Pennsylvania

218. Great Sacandaga Lake is located in what populous state where Montauk Point and Staten Island can be found?
New York

219. Cape Henlopen and Pea Patch Island are located in Delaware Bay and belong to what state?
Delaware

220. Lake Candlewood and the Hockanum River are located in what state bordering Long Island Sound to the south and whose capital is Hartford, the insurance capital of the United States?
Connecticut

221. What city is Connecticut is the home of the U.S. Naval Submarine Base?
Groton

222. The Punkin Chunkin World Championship is located in the city of Bridgeville in what state?
Delaware

223. Eartha, a model of our planet Earth, holds the world record as the world's largest rotating globe. Eartha is on display at Yarmouth, a city in what state?
Maine

224. Camp David is located in Catoctin Mountain Park, in what state bordering Virginia?
Maryland

225. Cranberries are the largest agricultural crop in what state that is home to the United States' first lighthouse, built in 1716?
Massachusetts

226. Moose are found throughout what state known by its nickname "The Granite State"?
New Hampshire

227. What state, whose largest city by population is Newark, is a leading producer of fresh fruits and vegetables?
New Jersey

228. Fire Island National Seashore is located in what state home to the Finger Lakes?
New York

229. The Liberty Bell is located in Philadelphia, the most populous city in what state?
Pennsylvania

230. Wild coyotes live on the islands of Narragansett Bay in what state bordering Mount Hope Bay?
Rhode Island

231. What state, bordering Lake Champlain, is the country's leading producer of maple syrup, ahead of New York?
Vermont

232. The Marshall Space Flight Center is located in Huntsville, a city in what state bordering Mobile Bay?
Alabama

233. The White River National Wildlife Refuge is located in what state with the city of Bentonville, where Walmart was founded?
Arkansas

234. Citrus fruits, such as oranges and grapefruits, make what state the leading producer of citrus in the country?
Florida

235. The Hartsfield-Jackson International Airport is located in Atlanta, the largest city in what state?
Georgia

236. The Mammoth Cave System is located in what state where "Derby Pie", a chocolate and walnut pastry, is a popular dessert?
Kentucky

237. What state bordering Chandeleur Sound to the east is located along the Gulf of Mexico?
Louisiana

238. The Marine Life Oceanarium is located in Gulfport in what state?
Mississippi

239. Uwharrie National Forest and Croatan National Forest are located in what state bordering Onslow Bay to the southeast?
North Carolina

240. Congaree National Park is located in what state where the Ashley and Cooper Rivers join at Charleston before flowing into the Atlantic Ocean?
South Carolina

241. Great Smoky Mountains National Park is located on the border between North Carolina and what other state whose most populous city is Memphis?
Tennessee

242. The Chesapeake Bay Bridge-Tunnel is located in what state home to the Luray Caverns?
Virginia

243. Monongahela National Forest is located in what state famous for its coal mining and Bridge Day, held annually in October?

West Virginia

244. The National Lincoln Monument is located in what state that ranks first in the United States for production of nuclear power?
Illinois

245. The cites of Evansville and Fort Wayne are located in what state bordering Lake Michigan to the north?
Indiana

246. Lexington, known as the "Horse Capital of the World", is one of the largest cities in what state in the United States?
Kentucky

247. Central Park and Times Square are located in what major city in the United States?
New York City

248. Hollywood and Disneyland are located in what major city in southern California?
Los Angeles

249. The Sears Tower, the second tallest tower in the United States, is located in what city in Illinois?
Chicago

250. What city named after Sam Houston is the largest city in Texas, before Dallas?

Houston

251. The Liberty Bell is located in what major city in Pennsylvania home to Independence Hall?
Philadelphia

252. What mountain range spans the west-central part of the United States?
Rocky Mountains

253. What mountain range is the dominant range in the eastern part of the United States?
Appalachian Mountains

254. The Sierra Nevada Mountains are located along the coast of what U.S. state?
California

255. The Cascade Range is located on the coast of Oregon and what other U.S. state?
Washington

256. The Brooks Range is located in the northern region of what large U.S. state?
Alaska

257. The highest peak in the United States, Denali, is located in what mountain range?
Alaska Range

258. What desert located on the border between Arizona and Mexico is known for the Saguaro Cactus?
Sonoran Desert

259. What desert, the largest in the United States, is located between the Sierra Nevada and the Wasatch Range?
Great Basin Desert

260. The Mojave Desert is located in the southern part of what U.S. state whose capital is Sacramento?
California

261. The Great Basin is a heart-shaped basin in the central part of what country bordering Canada and Mexico?
United States

262. What valley in southern California is the lowest point in the United States, and borders the Mojave Desert?
Death Valley

263. Hilo, one of the most populous cities in Hawaii, is the largest city on what island in Hawaii?
Hawaii

264. Kodiak Island is a large island off the southern coast of what U.S. state whose capital is Juneau?
Alaska

265. What island is a territory of the United States and is located in the Greater Antilles region of the Caribbean Sea?

Puerto Rico

266. What archipelago south of Alaska lies very close to the Queen Charlotte Islands?
Alexander Archipelago

267. What chain of islands in southwestern Alaska are home to the Andreanof Islands?
Aleutian Islands

268. Mount St. Helens is located in the southern part of what northwestern U.S. state whose capital is Olympia?
Washington

269. Lassen Peak is a famous mountain in the United States located in the northern part of what U.S. state?
California

270. Mauna Kea, the highest mountain in the world measured from the bottom of the sea, is located in what U.S. state?
Hawaii

271. Mauna Loa, which erupted in 1984, is a major mountain and volcano in what U.S. state?
Hawaii

272. The source of the Mississippi River is located in what lake in the northern part of Minnesota?
Lake Itasca

273. St. Louis and Minneapolis are major cities on what major river in the United States?
Mississippi River

274. The Colorado River's source is in La Poudre Pass Lake and its mouth is in what gulf south of Mexico?
Gulf of California

275. The source of the Snake River is located in what major mountain range in the west-central United States?
Rocky Mountains

276. The mouth of the Ohio River is located in what major river that is one of the longest in the world?
Mississippi River

277. Yuma is a city on what major river emptying out into the Gulf of California that is mostly in the United States?
Colorado River

278. Cincinnati and Pittsburgh are major cities on what river that is a right tributary of the Mississippi River?
Ohio River

279. What lake in the Great Lakes is the largest by area in the United States and the largest freshwater lake by area in the world?
Lake Superior

280. What lake, part of the five Great Lakes in the northern and eastern United States, is the smallest by area out of all of them?
Lake Ontario

281. Chesapeake Bay, a large bay on the coast of the northeastern United States, empties out into what ocean?
Atlantic Ocean

282. The Gulf of Mexico, south of the southeastern United States, borders how many U.S. states in total?
Five

283. The Dixon Entrance is a large strait located between Haida Gwaii and what archipelago in southern Alaska?
Alexander Archipelago

284. The world's largest roller coaster is located in what state in the Great Lakes region?
Ohio

285. What major city in Tennessee is famously known as the Music Capital of the World?
Nashville

286. What country in North America is known for having the most jobs in the world?
United States

287. What state in the northwestern United States has three times more cattle than people?
Montana

288. Montgomery is located north of the Piedmont Plateau in what state?
Alabama

289. Wheeler Lake is located on what river in Alabama's north?
Tennessee River

290. Little River Canyon National Park is located southeast of what Alabama city in the Appalachian Mountains?
Fort Payne

291. What major city in Alabama is located on the Black Warrior River?
Tuscaloosa

292. Eufaula, located on the Walter F. George Reservoir, is a city in what state?
Alabama

Canada, Greenland, and Mexico
Canada, Mexico, Greenland, St. Pierre and Miquelon

A Competitor's Compendium to the Geography Bee

1. Reservoir Manicouagan is located in what country?
   Canada

2. What country borders the Gulf of California?
   Mexico

3. The Hayes Peninsula is located in what territory?
   Greenland

4. Nuuk is the capital of what North American territory belonging to Denmark?
   Greenland

5. St. Pierre and Miquelon is an overseas territory of what country?
   France

6. What territory is located south of the Canadian island of Newfoundland?
   St. Pierre and Miquelon

7. Monterrey is a major city miles south of the Rio Grande in what country?
   Mexico

8. The Yucatan Peninsula and Eugenia Point are located in what country?
   Mexico

9. Vancouver Island belongs to what Canadian province?
   British Columbia

10. What Mexican state is the largest by area?
    Chihuahua

11. The Tropic of Cancer passes through what country – Mexico or Canada?
    Mexico

12. Godthab is another name for what territorial capital city?
    Nuuk

13. Kalaallit Nunaat is another name for what territory?
    Greenland

14. Greenland is located northeast of what country?
    Canada

15. Iceland is located southeast of what territory?
    Greenland

16. Ungava Bay borders the Hudson Strait in what country?
    Canada

17. The Nares Strait separates Baffin Bay from what sea?
    Lincoln Sea

18. Guadalajara is a major city in what country?
    Mexico

19. The Kicking Horse River, known for its roaring rapids, is located in the Rocky Mountains in what Canadian province?
    British Columbia

20. The village of Augpilagtoq is located in what territory?
Greenland

21. Chateau Frontenac can be found in what French-speaking city?
Quebec City

22. The Calgary Stampede is held in what Canadian province?
Alberta

23. What country ranks second in world uranium production?
Canada

24. What city in Canada is the second most populous French-speaking city in the world after Paris in France?
Montreal

25. Mexico takes its name from Mexica, another name for what major civilization in Mexico and Central America?
Aztec

26. Chicxulub is on the Yucatan Peninsula in what country?
Mexico

27. Lake Winnipegosis is located in what Canadian province?
Manitoba

28. Villahermosa is a city in Tabasco in what country bordering the United States?
Mexico

29. The Vizcaino Desert is on what peninsula in Mexico?

Baja California

30. The Kane Basin and Smith Bay border what territory?
Greenland

31. Scammon Lagoon is located in what country?
Mexico

32. The Isthmus of Chignecto is located in what Canadian province along with New Brunswick?
Nova Scotia

33. Knud Rasmussen Land is located in what territory?
Greenland

34. Steenstrup Glacier and Cape Cort Adelaer are located in what territory?
Greenland

35. The Long Range Mountains are in what province partly made up of the Avalon Peninsula?
Newfoundland and Labrador

36. Belcher Island is located in what bay bordering Manitoba and Nunavut?
Hudson Bay

37. Banff National Park is located in what province straddling the North Saskatchewan and Peace Rivers?
Alberta

38. The Sierra Madre del Sur is located in what country whose major cities include Netzahualcoyotl and Puebla?
Mexico

39. Xcaret Park is a site on what major peninsula bordering the Caribbean Sea?
Yucatan Peninsula

40. Eugenia Point is located on what major peninsula with the peak of Picacho del Diablo?
Baja California

41. Guadalajara is a city in what country bordering the Rio Bravo del Norte?
Mexico

42. El Chichon is a peak in what country?
Mexico

43. Mexico borders what major bay south of the Gulf of Mexico?
Bay of Campeche

44. Madre Lagoon is located in what country home to the Conchos River?
Mexico

45. Nuevo Leon is a state bordering Coahuila and Tamaulipas in what country?
Mexico

46. Oaxaca borders what state to the west?

Guerrero

47. Yucatan, Campeche, and Quintana Roo are states on what peninsula?
    Yucatan Peninsula

48. Sonora is west of what state with the major city of Juarez?
    Chihuahua

49. Ciudad Victoria is a city in what state?
    Tamaulipas

50. San Jose Island and San Margarita Island are off the coast of what state?
    Baja California Sur

51. Tabasco borders what bay to the north?
    Bay of Campeche

52. Oaxaca and Chiapas border what bay to the south?
    Gulf of Tehuantepec

53. The Angostura Reservoir is located in what state bordering Tabasco?
    Chiapas

54. Guadalajara is one of the most populous cities in Mexico. This city is located in what state home to Lake Chapala?
    Jalisco

55. The Sea of Cortes is also known by what name?
    Gulf of California

56. Durango and Sinaloa are states south of what state?
Chihuahua

57. Cape Corrientes borders what bay to the north?
Banderas Bay

58. The Kapiskau River is located in what country where Lake Nipigon can be found?
Canada

59. Cape Corrientes and Cape Rojo can be found in what mountainous country?
Mexico

60. Choctawhatchee Bay and Pensacola Bay are in what gulf?
Gulf of Mexico

61. Nunivak Island is separated from what state by the Etolin Strait?
Alaska

62. The Rae Isthmus and the Melville Peninsula can be found in what Canadian territory bordering Hudson Bay?
Nunavut

63. The Horn Plateau is in the southern region of what Canadian territory?
Northwest Territories

64. Prince Albert National Park is in what province bordering North Dakota and Montana?

Saskatchewan

65. Pukaskwa National Park can be found in what province known for its Manitoulin Island?
Ontario

66. Wood Buffalo National Park is located in the Northwest Territories and what province?
Alberta

67. The Ogilvie Mountains are located in what territory whose capital is Whitehorse?
Yukon

68. The Smallwood Reservoir is located in what province bordering the Strait of Belle Isle?
Newfoundland and Labrador

69. Reindeer Lake is located in Saskatchewan and what province?
Manitoba

70. What lake is the largest in Manitoba?
Lake Winnipeg

71. Lesser Slave Lake is located in what province whose capital is Edmonton and contains the cities of Calgary and Fort Saskatchewan?
Alberta

72. Southampton Island and Coats Island are located in what Canadian territory that is home to Ukkusiksalik National Park?
Nunavut

73. Mississauga and Brampton are cities in what province bordering Georgian Bay?
Ontario

74. Akimiski Island in James Bay belongs to what Canadian territory?
Nunavut

75. St. Pierre and Miquelon is west of what major peninsula on the island of Newfoundland?
Burin Peninsula

76. Charlottetown is the capital of what province bordering Egmont Bay and Rollo Bay?
Prince Edward Island

77. Chignecto Bay is an inlet of what major bay?
Bay of Fundy

78. The Notre Dame Mountains are in what province bordering New Brunswick?
Quebec

79. The Laurentian Mountains are in the southern region of what province bordering Ontario?
Quebec

80. Mount Royal is a small mountain in what city in Quebec?
Montreal

81. Lake Chapala is located in the southwestern region of what country with the cities of Leon and Monterrey?
Mexico

82. Winnipeg, the capital and largest city in Manitoba, is located at the confluence of the Red and what other river?
Assiniboine River

83. The Isthmus of Tehuantepec is located in what country home to Terminos Lagoon?
Mexico

84. Cozumel is an island belonging to what country bordering the United States?
Mexico

85. What island, the fifteenth largest in Canada, is separated from the Boothia Peninsula by the James Ross Strait?
King William Strait

86. The Wernecke Mountains and Ogilvie Mountains are located in what territory?
Yukon

87. The sites of Xpuhil and Calakmul are located in what Mexican state bordering Quintana Roo?
Campeche

88. Greenland borders what sea to the northeast that is a major sea in the Arctic Ocean?
Greenland Sea

89. Riding Mountain National Park is located south of Dauphin Lake in what province?
Manitoba

90. Coats Island and Mansel Island belong to what territory?
Nunavut

91. The Stikine Mountain Ranges are located in what province bordering Queen Charlotte Sound and the Hecate Strait?
British Columbia

92. Nahanni National Park is located in what territory whose capital is Yellowknife?
Northwest Territories

93. Lake Simcoe is located in what province bordering Nottawasaga Bay?
Ontario

94. Pukaskwa National Park is located in what province bordering James Bay?
Ontario

95. Miramichi Bay and Tabusintac Bay border what province whose capital is Fredericton?
New Brunswick

96. The Northumberland Strait separates what province from New Brunswick and Nova Scotia?
Prince Edward Island

97. Gros Morne National Park is located in the southern region of what province?
Newfoundland and Labrador

98. Shibogama Lake and Lake Nipigon are located in what province bordering Nipigon Bay?
Ontario

99. Montreal is the second most populous French-speaking city in the world and is located in what province?
Quebec

100. Toronto is the most populous city in what province bordering Lake Ontario, Lake Superior, Lake Huron, and Lake Erie?
Ontario

101. There is a sculpture of an inverted country church in what major city in British Columbia, in Canada?
Vancouver

102. What country in North America is known as the "Land of Maples" and the "Land of Lilies"?
Canada

103. The region of southern Ontario is known as "The Golden Horseshoe" and is located in what country?
Canada

104. What major city in Ontario and Canada is the most populous city in the Great Lakes region?
Toronto

105. Toronto, which is one of the most diverse cities in the world, was formerly known as York and is located in what country?
Canada

106. What major city in Canada was previously known as "Ville-Marie" and is a chief port city on the St. Lawrence River?
Montreal

107. What city, the most densely populated in Canada, is located on the Strait of George and the Buzzard Inlet?
Vancouver

108. Ottawa, which means "to trade" in the Algonquian language of the Native Americans, is located on what river?
Ottawa River

109. Calgary, located on the Bow River, is the fifth largest city in Canada and the largest city in what Canadian province?
Alberta

110. The highest peak in Canada is what peak located in the St. Elias Mountains in Canada?
Mount Logan

111. The Rocky Mountains, located in the United States, also extend across nearly the entire western part of what country?
Canada

112. The Coast Mountains, part of the Pacific Coast Ranges, extend across the Alaskan Panhandle, Yukon, and what other Canadian province?
British Columbia

113. The Cascade Range, part of the Pacific Ocean's Ring of Fire, is located in the northwestern United States and what Canadian province?
British Columbia

114. The Brooks Range, also known as the Arctic Mountains, are located in Alaska and extend into what Canadian territory?
Yukon

115. The St. Elias Mountains, a subrange of the Pacific Coast Ranges, are home to what peak that is the highest in Canada?
Mount Logan

116. The Osoyoos Desert, which has the same name as a city in Canada, is located in what Canadian province?
British Columbia

117. The Mackenzie River Basin, which borders the Arctic Ocean to the north, is located in what administrative division of Canada?
Northwest Territories

118. The Fraser Valley is located in the southwestern region of what Canadian province whose capital is Victoria?
British Columbia

119. The Ottawa Valley, formed by the Ottawa River, is located in Ontario and what other Canadian province?
Quebec

120. Baffin Island, the largest island in Canada, borders Baffin Bay, the Davis Strait, and the Hudson Strait in what Canadian territory?
Nunavut

121. Ellesmere Island, an island in the Queen Elizabeth Islands in northern Canada, is located in what Canadian territory?
Nunavut

122. What major city is located on Vancouver Island, in the southwestern region of British Columbia in Canada?
Victoria

123. The Canadian Arctic Archipelago, located north of the Canadian mainland, consists of islands in the Northwest Territories and what Canadian territory?
Nunavut

124. Mount Edziza, the highest volcano in Canada is located in the northwestern region of British Columbia near what well preserved black cinder cone?
Eve Cone

125. The Columbia River, whose mouth is in the Pacific Ocean, has its source in what lake?
Columbia Lake

126. Vancouver and Revelstoke are major cities on what Canadian river that empties out into the Pacific Ocean?
Columbia River

127. The Fraser River, whose mouth is in the Strait of George, has its source in what mountain pass?
Fraser Pass

128. Fort Providence is a city on what river whose source is in the Great Slave Lake in Canada?
Mackenzie River

129. The source of the Ottawa River is located in what lake in southern Canada?
Lake Capimitchigama

130. Montreal and Quebec City are located in what river that empties out into the Gulf of Saint Lawrence?
Saint Lawrence River

131. Mount Royal is a triple-peaked mountain overlooking what city in the Quebec province?
Montreal

132. What city in western Canada and British Columbia is the most densely populated city in Canada?
Vancouver

133. What city is the fourth most populous city in North America and the most populous city in Canada?
Toronto

134. What territory in Canada is the country's largest administrative division by area?
Nunavut

135. What province in Canada is the smallest by area?
Prince Edward Island

136. The mouth of the Nelson River is located in what major bay in eastern Canada?
Hudson Bay

137. Nipawin is a major city on what river that empties out into the Nelson River?
Saskatchewan

138. The South Saskatchewan River is formed by the confluence of the Oldman River and what other river that flows through Calgary, Alberta?
Bow River

139. Lake Athabasca is located in the northwestern corner of Saskatchewan and is shared with what other Canadian province?
Alberta

140. Akimiski Island, located in James Bay, belongs to what Canadian territory?
Nunavut

141. Akpatok Island is situated in the northern part of what bay?
Ungava Bay

142. The Dease Strait separates Kent Peninsula from what major island?
Victoria Island

143. Mansel Island is located off the coast of what major peninsula bordering the Hudson Strait?
Ungava Island

144. Anticosti Island, located south of the Laurentide Scarp, is in what gulf?
Gulf of St. Lawrence

145. Bras d'Or Lake is located on what island bordering the Strait of Canso?
Cape Breton Island

146. The Belcher Islands are located south of the King George Islands in what bay?
Hudson Bay

147. The Dundas Peninsula, which borders Viscount Melville Sound, is located on what island?
Melville Island

148. The Foxe Channel separates the Foxe Peninsula on Baffin Island from the Bell Peninsula on what island?
Southampton Island

149. The Kennedy Channel separates Ellesmere Island from what territory of Denmark?
Greenland

150. The Chesterfield Inlet stretches into what Canadian territory?
Nunavut

151. The Brodeur Peninsula is at the western end of what island that is the largest in Canada by area?
Baffin Island

152. Cape Breton Island is separated from the island of Newfoundland by what strait connecting the Gulf of Saint Lawrence to the Atlantic Ocean?
Cabot Strait

153. Cumberland Sound, which empties out into the Labrador Sea and Davis Strait, borders which major island?
Baffin Island

154. Lancaster Sound separates Baffin Island from what island to the north?
Devon Island

155. Cape Parry, which borders Franklin Bay to the west, borders what gulf to the north that borders Banks Island?
Amundsen Gulf

156. The Collinson Peninsula and Wollaston Peninsula are located on what island bordering Viscount Melville Sound to the north?
Victoria Island

157. The Grinnell Peninsula is located on what island bordering Baffin Bay, Jones Sound, and Lancaster Sound?
Devon Island

158. The Fraser Plateau is located west of the Rocky Mountains and east of what major mountain range?
Coast Mountains

159. The Adelaide Peninsula borders what gulf to the west?
Queen Maud Land

160. The Yukon Plateau, which borders the Ogilvie Mountains to the northwest and the Selwyn Mountains to the east, is located in what Canadian territory?
Yukon

161. Haida Gwaii, which is separated from mainland Canada by the Hecate Strait, is known more commonly by what English name?
Queen Charlotte Islands

162. The Gulf of Boothia separates the Simpson Peninsula from what large island to the north situated north of the Foxe Basin and east of the Prince Regent Inlet?
Baffin Island

163. The Great Plain of the Koukdjuak is located north of Amadjuak Lake and southwest of Nettilling Lake on what island?
Baffin Island

164. Lake Winnipeg and Southern Indian Lake are major lakes in what Canadian Province bordering Hudson Bay to the northeast?
Manitoba

165. Great Bear Lake and Great Slave Lake, two of the largest lakes by area in Canada, are both located in what Canadian territory?
Northwest Territories

166. The Mackenzie Mountains extend over the Northwest Territory and what other administrative division?
Yukon

167. Prince Albert and Moose Jaw are major cities in what Canadian province whose capital and largest city is Regina?
Saskatchewan

168. The Cassiar Mountains extend from Yukon south into what Canadian province bordering Alaska to the west and Washington, Idaho, and Montana to the south?
British Columbia

169. Calgary is one of the most populous cities in what country bordering the Beaufort Sea and the Labrador Sea?
Canada

170. Lake Nipigon is north of the Trans-Canada Highway System in what southern Canadian province whose capital is Toronto?
Ontario

171. The CN Tower, the highest tower in Canada, attracts two million international visitors annually and is located in what major city?
Toronto

172. The Canadian Rocky Mountain Parks are a UNESCO World Heritage Site in British Columbia and what other province?
Alberta

173. The Scarborough Bluffs, located on the shoreline of Lake Ontario, are in what major city?
Toronto

Central America
Guatemala, Nicaragua, Honduras, Panama, Belize, Costa Rica, El Salvador

1. Cordillera de Talamanca is in Costa Rica and what other country?
   Panama

2. Volcan Tajmulco is a high mountain in the Sierra Madre. It is the highest peak in what country?
   Guatemala

3. Lake Bayano is located east of Gatun Lake in what country bordering Colombia?
   Panama

4. Golfo de Fonseca forms the southern coast of what country in Central America?
   Honduras

5. Cerro El Pital is on a country's border with Honduras. Name this country, bordering Jiquilisco Bay.
   El Salvador

6. Islas de la Bahia belongs to what country?
   Honduras

7. Coral colonies growing along much of the coast of Belize form the longest barrier reef in which hemisphere – east or west?
   West

8. Lake Izabal is located in what country is located in what country containing the Sierra de las Minas in the south?
   Guatemala

9. The Maya Mountains are located in what country with the Turneffe Islands?
   Belize

10. The Gulf of Honduras borders what country to the southwest?
    Guatemala

11. Nueva San Salvador is a city west of San Salvador in what country?
    El Salvador

12. Golfo de Nicoya borders what major peninsula in Costa Rica?
    Nicoya Peninsula

13. Punta Roca is one of the most popular beaches in what country bordering the Gulf of Fonseca?
    El Salvador

14. The Cordillera Entre Rios is south of the Patuca River in what country containing Cusuco National Park?
    Honduras

15. Estadio Cuscatlan is the largest stadium in Central America, and can be found in what country bordering Honduras?
El Salvador

16. Isla de Guanaja belongs to what country containing the Sico and Sutaco Rivers?
Honduras

17. Bosawas Biosphere Reserve is part of what country bordering Salinas Bay to the west?
Nicaragua

18. The Cordillera de Talamanca is located in what country whose capital is San Jose?
Costa Rica

19. The Catedral de Leon is the largest cathedral in Central America, in what country?
Nicaragua

20. Coronado Bay borders what country home to Monteverde Forest Reserve?
Costa Rica

21. Tortuguero National Park is located on the Caribbean coast of what country?
Costa Rica

22. David and La Chorrera are major cities in a country that is also an isthmus. Name this countr?
Panama

23. The Gulf of Chiriqui borders what country containing the Chucunaque and Tuira Rivers?
    Panama

24. Coiba National Park is located in what country containing the Azuero Peninsula?
    Panama

25. Mosquito Gulf borders what country containing Coiba Island?
    Panama

26. Lake Izabal is located in what country bordering Honduras?
    Guatemala

27. What lake is the largest in Central America?
    Lake Nicaragua

28. What country borders Coronado Bay to the south?
    Costa Rica

29. Tortuguero National Park is located in what country?
    Costa Rica

30. Volcan Masaya National Park is located in what country?
    Nicaragua

31. Caratasca Lagoon is located in what country bordering Nicaragua?
    Honduras

32. Cape Santa Elena is located in what country?
    Costa Rica

33. What English-speaking Central American country was formerly known as British Honduras?
    Belize

34. The Valiente Peninsula is located in what country bordering the Gulf of Uraba?
    Panama

35. David and Dolega are cities in the western region of what country bordering Costa Rica?
    Panama

36. Las Minas is the highest point in what country bordering the Gulf of Honduras?
    Honduras

37. Guanaja Island and Utila Island, part of the Bay Islands, are located north of what country?
    Honduras

38. What country is home to the Cordillera Isabelia in the north?
    Nicaragua

39. Jinotega and Chontales are departments of what country?
    Nicaragua

40. Cape Camaron is located in what country bordering El Salvador and Guatemala?

Honduras

41. Puerto Cabezas and Bluefields are cities in what country bordering Honduras?
Nicaragua

42. How many countries border Guatemala?
Four

43. Arenal Volcano is located in the province of Alajuela in what country?
Costa Rica

44. Copan is an archaeological site of the Mayan civilization in what country?
Honduras

45. The Corn Islands is one of the twelve municipalities of the South Caribbean Coast Autonomous Region. These islands belong to what country?
Nicaragua

46. Altun Ha is the name of the ruins of a Mayan city in the northern part of what country?
Belize

47. Volcan de Fuego is an active stratovolcano in what country whose currency is the quetzal?
Guatemala

48. Punta Mariato is the southernmost point on the North American mainland. This cape is located in the Veraguas province on the Azuero Peninsula in what country?
Panama

49. El Imposible National Park is located in the Ahuachapan Department in what country bordering the Pacific Ocean?
El Salvador

50. What city, the second most populous in Central America after Guatemala City, is located on the southwestern shore of Lake Managua?
Managua

51. Atlitan Nature Reserve is located in what country bordering the Gulf of Honduras?
Guatemala

52. The Midwinters Lagoon is located in what country bordering Chetumal Bay to the north?
Belize

53. Cerro Cacahuatique is a peak in what country bordering the Gulf of Fonesca?
El Salvador

54. The Bay Islands belong to what country where Caratasca Lagoon can be found?
Honduras

55. Mono Point is located in what country where the cities of Chinandega and Masaya are located?

Nicaragua

56. The Nicoya Peninsula is the largest peninsula in what country bordering Coronado Bay?
Costa Rica

57. Chiriqui Lagoon is located in what country bordering the Gulf of Chiriqui?
Panama

58. The Gulf of Mosquitos borders what country where Isla de Coiba National Park can be found?
Panama

59. The Monteverde Forest Reserve and the Osa Peninsula are located in what country bordering Dulce Gulf?
Costa Rica

60. Burica Point is located on the Burica Peninsula in what country whose highest point is Cerro Chirripo at 12,530 feet?
Costa Rica

61. The Bosawas Biosphere Reserve is located in what country where the active Masaya Volcano can be found?
Nicaragua

62. Cusuco National Park is located in what country bordering the Caribbean Sea to the north?
Costa Rica

63. Estadio Cuscatlan, the largest stadium in Central America, is located in what country?
El Salvador

64. The Maya Mountains are located in what country where the Blue Hole National Monument and the Turneffe Islands can be found?
Belize

65. Lake Izabal is located in what country where Tikal National Park can be found in the country's northern region?
Guatemala

The Caribbean

Cuba, Haiti, Dominican Republic, Jamaica, Puerto Rico, Bahamas, Trinidad and Tobago, Dominica, St. Lucia, Barbados, St. Vincent and the Grenadines, St. Kitts and Nevis, Antigua and Barbuda, U.S. Virgin Islands, British Virgin Islands, Anguilla, Cayman Islands, Bermuda, Aruba, Bonaire, Curacao, Saba, Montserrat, Guadeloupe, Martinique, St. Barthelemy, St. Martin, Turks and Caicos Islands, Navassa Island

1. What country is the largest in the Caribbean?
Caribbean

2. Haiti and what other country make up the island of Hispaniola?
Dominican Republic

3. Puerto Rico is a territory of what country?
United States

4. What is the capital of Jamaica?
Kingston

5. Aruba is part of what Antilles?
Netherlands Antilles

6. Curacao belongs to what country?
Netherlands

7. Aruba, Bonaire, and Curacao, part of the Netherlands Antilles, are known by what name?
ABC Islands

8. The Virgin Islands are split between what two countries?
United States and United Kingdom

9. Which island is larger – Antigua or Barbuda?
Antigua

10. The Archipielago de Sabana in Cuba borders what channel?
Nicholas Channel

11. The Bellamar Caves are located in what country bordering the Gulf of Batabano?

Cuba

12. What is the capital of Curacao?
    Willemstad

13. Bequia Island belongs to what country separates from Saint Lucia by the St. Vincent Passage?
    St. Vincent and the Grenadines

14. The Gulf of Paria borders Venezuela and what Caribbean country?
    Trinidad and Tobago

15. What country is the largest and the most populous in the Lesser Antilles?
    Trinidad and Tobago

16. Montserrat is a territory of what country in Europe?
    United Kingdom

17. St. Eustatius and Saba are territories in the Caribbean Sea belonging to what country?
    Netherlands

18. The Anegada Passage is west of what territory of the United Kingdom whose capital is The Valley?
    Anguilla

19. Fort-de-France is the capital of what French territory separated from Dominica by the Martinique Passage?
    Martinique

20. Charlotte Amalie is the capital of what U.S. Territory east of Puerto Rico?
U.S. Virgin Islands

21. Montego Bay borders what country containing Portland Point and Morant Point?
Jamaica

22. Spanish Town is a city west of Kingston, the capital of what country whose population is a little over 2.7 million?
Jamaica

23. Gonave Island belongs to what French-speaking country bordering the Windward Passage?
Haiti

24. Cueva de las Maravillas is a site in what country whose capital is Santo Domingo?
Dominican Republic

25. The Cordillera Central is located in the western part of what country bordering Bahia de Ocoa?
Dominican Republic

26. Grand Bahama Island is located in what country containing Lake Rosa?
Bahamas

27. Paradise Island is a site in what country containing Lucayan National Park?
Bahamas

28. Port-au-Prince is the capital of what country bordering Manzanillo Bay?
Haiti

29. Great Abaco Island belongs to what country whose capital is Nassau?
Bahamas

30. Puerto Rico is separated from the island of Hispaniola by what passage?
Mona Passage

31. The Cayman Islands are south of what country bordering Batabano Gulf?
Cuba

32. The Yucatan Channel separates Cuba from what country?
Mexico

33. Grande-Terre and Marie-Galante are islands in what French territory?
Guadeloupe

34. Christoffel Park is located on what island containing the Hato Caves?
Curacao

35. Harrison's Cave is located in what country whose capital is Bridgetown?
Barbados

36. Cienfuegos is the capital of what Cuban province bordering Villa Clara?
Cienfuegos

37. Pigeon Island National Landmark can be found in what country?
St. Lucia

38. Le Lamentin is a city in what French territory bordering the St. Lucia Channel and the Martinique Passage?
Martinique

39. Abaco National Park is located on the island of Abaco in what country west of the Turks and Caicos Islands?
Bahamas

40. Inagua National Park is located on what island in the Bahamas?
Great Inagua

41. Pico Cristal National Park can be found in the eastern part of what country bordering the Windward Passage?
Cuba

42. Sierra de Bahoruco National Park is located in what country containing Lake Enriquillo?
Dominican Republic

43. Piarco International Airport and the cities of Tanapuna and Arima can be found in what country bordering Guayaguayare Bay?
Trinidad and Tobago

44. In especially Florida and the West Indies, a small low-lying island usually made up of coral or sand is known by what term?
Key

45. The Sierra Maestra is located in what country bordering the Isle of Youth?
Cuba

46. Mount Gimie is the highest point in what country whose capital is Castries?
Saint Lucia

47. Boiling Lake, which can be found in Morne Trois Pitons National Park, is one of the world's largest thermal lakes. This lake is located in what small country in the Lesser Antilles?
Dominica

48. Isla de la Juventud, one of the 350 islands in the Canarreos Archipelago, is the second largest island in what country in the Greater Antilles?
Cuba

49. The florin is the official currency of what territory in the Netherlands Antilles?
Aruba

50. What is the second largest religion followed in Trinidad and Tobago after Christianity?
Hinduism

A Competitor's Compendium to the Geography Bee

South America

Northwestern South America
Colombia, Venezuela, Peru, Ecuador, Bolivia

1. Lake Maracaibo is in what country?
   Venezuela

2. Guayaquil is the largest city in what country?
   Ecuador

3. Lima is a port city on the Pacific Ocean in what country?
   Peru

4. The Maranon River can be found in what country bordering Ecuador?
   Peru

5. Medellin is a city in what country with 48 million people?
   Colombia

6. Chimborazo is a peak in what country?
   Ecuador

7. Colombia and Venezuela share a tropical grassland called what?
   Llanos

8. Barranquilla and Cartagena are port cities on the Caribbean Sea in what country?
   Colombia

9. The Gulf of Venezuela borders Venezuela and what country?
   Colombia

10. Angel Falls is located in what country?
    Venezuela

11. Cali is a major city in the western region of what country?
    Colombia

12. Machu Picchu is a historical site in what country?
    Peru

13. Lake Titicaca, the world's highest navigable lake, is shared by what two countries?
    Bolivia and Peru

14. The Apurimac and Ucayali Rivers are located in what country?
    Peru

15. Quito is the capital of what country that owns the Galapagos Islands?
    Ecuador

16. Armenia is a city in what country?
    Colombia

17. Santa Marta is a city on the northern coast of what country?
Colombia

18. Ciudad Guayana is a city on the Orinoco River in what country?
Venezuela

19. The Gulf of Paria borders what South American country?
Venezuela

20. Rogaguado Lagoon is located in what country?
Bolivia

21. A huge statue of Jesus Christ stands on San Pedro Hill in what Bolivian city?
Cochabamba

22. The Cordillera Vilcabamba is located in what country?
Peru

23. The Gulf of Guayaquil is named after what Ecuadorian city?
Guayaquil

24. The Gold Museum, displaying shiny objects made by the natives of South America before the Europeans, can be found in Bogota in what country bordering Docampado Bay?
Colombia

25. Galera Point is found in what country with the cities of Riobamba and Machala?
Ecuador

26. Puerto Ayora is in the Galapagos Islands in what ocean?
Pacific Ocean

27. El Avila National Park is a famous attraction in what country bordering Serpent's Mouth?
Venezuela

28. Tortumo Volcano is a famous tourist attraction in Santa Catalina in what country?
Colombia

29. Trujillo and Chiclayo are cities on the coast of what country?
Peru

30. Nevado Huascaran is the highest point in what country?
Peru

31. Bolivia is landlocked, surrounded by Brazil, Chile, Argentina, Paraguay, and what country?
Peru

32. The Gate of the Sun, a structure done by the people of the Inca Empire, is a doorway cut from a huge stone with bird-humanlike figures carved into it. This doorway is located in Tiwanaku in what country where the Parapeti River can be found?
Bolivia

33. Mochima National Park and El Avila National Park are located in what country containing the Orinoco River?
Venezuela

34. Machu Picchu is an ancient Inca site in what country containing the cities of Huancayo and Chimbote?
Peru

35. Biocentro Guembe is a natural paradise in what country whose major cities include Oruro and Cochabamba?
Bolivia

36. Uyuni Salt Flat is the world's largest inland area of salt, located in what country containing Lake Rogagua?
Bolivia

37. Lake Titicaca is shared by Bolivia and what country containing Point Parinas and Point Aguja?
Peru

38. The Orinoco River is primarily located in what country whose capital is Caracas?
Venezuela

39. Mt. Chimborazo is in the Andes Mountains in what country bordering the Gulf of Guayaquil?
Ecuador

40. The Paraguana Peninsula is in what country bordering the Caribbean Sea?
Venezuela

A Competitor's Compendium to the Geography Bee

41. Manizales and Monteria are cities in what country bordering Peru to the south?
Colombia

42. The Ucayali River is in what country whose capital is Lima?
Peru

43. The Orinoco River Delta is located in what country?
Venezuela

44. The Desaguadero River is located in what country containing Lake Rogagua?
Bolivia

45. Iquitos is a major city in Loreto in what country bordering Ecuador?
Peru

46. Barrancabermeja, Piedecuesta, and Bucaramanga are cities in what country bordering the Gulf of Morrosquillo?
Colombia

47. The Galapagos Islands are a volcanic chain of islands in the Pacific Ocean belonging to what country?
Ecuador

48. Sechura Desert is located in the northwestern region of what country?
Peru

49. The Gulf of Darien borders Panama and what South American country?
Colombia

50. Quilotoa is a water-filled caldera in what country bordering the Pacific Ocean?
Ecuador

51. The Beni River is a tributary of what river in Bolivia and Peru?
Madre de Dios River

52. What city in Ecuador is the economic and political center of the Portoviejo River Valley with a metropolitan population of approximately 285,000 people?
Portoviejo

53. The Meta River is a left tributary of what major river in Venezuela?
Orinoco River

54. Los Nevados National Natural Park is located in what branch of the Colombian Andes?
Cordillera Central

55. Iquitos is located in the northern region of what country that was home to the Norte Chico Civilization?
Peru

A Competitor's Compendium to the Geography Bee

Southern South America
Argentina, Chile, Paraguay, Uruguay, Falkland Islands, Saint Helena, Ascension, Tristan de Cunha

1. The Atacama Desert can be found in the northern region of what country?
   Chile

2. Tierra del Fuego is an island belonging to Chile and what other country?
   Argentina

3. The Rio de la Plata empties out into the Atlantic Ocean and borders Argentina and what country?
   Uruguay

4. Cerro Aconcagua is a mountain in what mountain range?
   Andes Mountains

5. The Parana River empties out into what ocean?
   Atlantic Ocean

6. Easter Island belongs to what country?
   Chile

7. The world's largest copper deposit is at Chuquicamata in what country?

Chile

8. Which country has a higher literacy rate – Uruguay or Paraguay?
   Uruguay

9. Guarani is the name of the people native to what country?
   Paraguay

10. What dam is the world's largest operating hydroelectric power plant?
    Itaipu Dam

11. The Cordillera de Paine is part of a national park that was made a world heritage site in 1978 in what country?
    Chile

12. Patagonia is a region in Chile and what country?
    Argentina

13. Guarani is spoken mainly in what country?
    Paraguay

14. What is the official language of Uruguay?
    Spanish

15. The Gran Chaco comprises parts of Argentina and what other country?
    Paraguay

16. Caaguazu is a department of what country whose capital is Asuncion?

A Competitor's Compendium to the Geography Bee

Paraguay

17. Oruro is a UNESCO cultural heritage site in what country with part of the Madre de Dios River in the north?
Bolivia

18. Spanish and what other language are spoken the most in South America?
Portuguese

19. Islas Malvinas is another name for what English territory east of Argentina with a little over 3,000 people?
Falkland Islands

20. Argentina and what country share the Rio de la Plata?
Uruguay

21. Los Glaciares National Park is located in what country whose currency is the peso?
Argentina

22. Camarones Bay and the Gulf of San Jorge border what country that contains the cities of Rosario and San Justo?
Argentina

23. Estadio Centenario is a stadium in what country containing the Cuchilla Grande and Lake Rincon del Bonete?
Uruguay

24. Ybycui National Park is located in what country containing the Tebicuary and Monte Lindo Rivers?
Paraguay

25. Chiloe Island belongs to what country home to the Atacama Desert?
Chile

26. Cape Tres Puntas and Cape San Antonio are located in what country bordering Uruguay?
Argentina

27. San Ambrosio Island belongs to what country?
Chile

28. Patagonia is in the southern region of Chile and what country bordering the Gulf of San Jorge?
Argentina

29. The Gulf of San Matias is north of what peninsula in Argentina?
Valdes Peninsula

30. Bustamante Bay is south of Cabo Dos Bahias in what country whose Baritu National Park borders the Tarija Province of Bolivia?
Argentina

31. Arauco Gulf and Inglesa Bay border what country?
Chile

32. The Gran Chaco extends across Paraguay and what country?
Argentina

33. The headquarters of MERCOSUR are located in what city in Uruguay?
Montevideo

34. Cordoba, Chaco, and Neuquen are provinces in what country?
Argentina

35. Santa Teresa National Park is a seaside forested area in the Rocha Department of what country?
Uruguay

36. San Jorge Gulf and Samborombon Bay border what country?
Argentina

37. Defensores Del Chaco National Park is located in what country?
Paraguay

38. The Taitao Peninsula and Isthmus of Ofqui are major landforms in what country whose highest point is the massive stratovolcano of Nevado Ojos del Salado, at 22,615 feet?
Chile

39. Lautaro Volcano is a peak in what country home to the Chonos Archipelago?

Chile

40. Alacalufes National Reserve is located between Canal Concepcion and the Strait of Magellan in what country?
Chile

41. The Magellanic subpolar forests cover parts of Chile and what other country?
Argentina

42. Concordia is a city in the Entre Rios province of what country bordering the San Matias Gulf?
Argentina

43. Ciudad del Este, the capital of the Alto Parana Department, is the second largest city in what country?
Paraguay

44. The Rio de la Plata is an estuary formed by the confluence of the Parana River and what other river?
Uruguay River

45. Rincon del Bonete Lake is the largest body of freshwater in what country?
Uruguay

46. Acaray Dam is a hydroelectric dam in Hernandarias, a city in what country?
Paraguay

47. Chiloe National Park, on the western coast of Chiloe Island, is located in what country?

Chile

48. General San Martin Park is located in the city of Mendoza in what country?
    Argentina

49. Spanish and Guarani are the official languages of what landlocked country?
    Paraguay

50. The headquarters of MERCOSUR are in what major city in Uruguay?
    Montevideo

Brazil and the Guianas
Brazil, Guyana, Suriname, French Guiana

1. What country in the Guianas was colonized by the Dutch?
   Suriname

2. Spaceport, a launch site of the European Space Agency, is in what overseas department of France in the Guianas?
   French Guiana

3. Marajo Island belongs to what country?
   Brazil

4. The Brazilian Highlands constitute part of what country?
   Brazil

5. Boundaries are claimed by Suriname in French Guiana and what country?
Guyana

6. Sugarloaf Mountain and Christ the Redeemer are major sites in what major city?
Brazil

7. Salvador, also known as Bahia, Is a major city on what country's eastern coast?
Brazil

8. What country has the highest literacy rate in the Guianas?
Guyana

9. The predominant languages in the Guianas are English, Dutch, French, and what other major language from India?
Hindi

10. What animal is the largest member of the cat family native to the Americas?
Jaguar

11. Iguazu Falls extends 2.5 miles along the border between Argentina and what country?
Brazil

12. Sugarloaf Mountain is a 1,300-foot tall block of granite in what city?
Rio de Janeiro

13. The Amazon village of Paragominas is located in what country?
Brazil

14. Samba is often called what country's official music?
Brazil

15. Sao Luis, in the state of Maranhao, is located on the Sao Marcos Bay in what country?
Brazil

16. Brownsberg Nature Park is situated near the Brokopondo Reservoir in what country?
Suriname

17. What is the official currency of French Guiana?
Euro

18. Bellevue de l'Inini is the highest point in what overseas department of France?
French Guiana

19. The Tapajos River is located in what country?
Brazil

20. The Mato Grosso Plateau is located in what country containing the Parnaiba River?
Brazil

21. Patos Lagoon is located in what country?
Brazil

22. The Xingu River is a tributary of what major South American river?
Amazon River

23. Cape Frio is located in what country containing part of the Guiana Highlands?
Brazil

24. Paulo Afonso Falls is located in what country bordering Paraguay?
Brazil

25. The Xingu and Madeira Rivers are major rivers in what country?
Brazil

26. Campinas is a city in what country containing the Paranaiba and Sao Francisco Rivers?
Brazil

27. The Araguaia River, a tributary of the Tocantins River, has its source in what mountain range?
Brazilian Highlands

28. Patos Lagoon is located in what country bordering Uruguay?
Brazil

29. Georgetown Football Stadium is located in the city of Georgetown in what country with a significantly high percentage of Hindus?
Guyana

30. Van Blommesterin Meer is a lake in what country bordering Guyana?
Suriname

31. Name the country where Mapinguari National Park can be found.
Brazil

32. Wonotobo Vallen is a lake in Suriname on the edge of what mountain range?
Guiana Highlands

33. Hinduism is one of the major religions followed by many people in what country bordering French Guiana to the direct east?
Suriname

34. French Guiana borders what small bay situated to its northeast?
Oyapock Bay

35. The Kanuku Mountains are located in what country bordering Suriname?
Guyana

36. Kayser Gebergte is a peak in the southern region of what country?
Suriname

37. French Guiana borders what country to the south?
Brazil

38. The Juruena and Purus Rivers are located in what country bordering the Atlantic Ocean to the east?
Brazil

39. The Courantyne River is the longest river in Suriname and originates in what mountain range?
Acarai Mountains

40. Fernando de Noronha is an archipelago of islands in the Atlantic Ocean. This UNESCO World Heritage Site is located off the coast of what country where the Jari River, a tributary of the Amazon River, can be found on the border between the states of Para and Amapa?
Brazil

A Competitor's Compendium to the Geography Bee

Asia

South Asia
India, Pakistan, Bangladesh, Nepal, Afghanistan, Sri Lanka, Bhutan, Maldives, and Myanmar

1. The Chennai Metropolitan area is the largest in what South Indian state?
   Tamil Nadu

2. Royal Chitwan National Park is in what country bordering Tibet?
   Nepal

3. China claims much of what Indian state?
   Arunachal Pradesh

4. Trincomalee is a major city in what South Asian island?
   Sri Lanka

5. Bhubaneswar is in what country bordering the Gulf of Kutch?
   India

6. Lucknow and Jaipur are major cities in what Indian state?
   Uttar Pradesh

7. Phuentsholing is a city in what country?
   Bhutan

8. What river delta is the largest in the world?
   Ganges River Delta

9. The Karnaphuli Reservoir is situated in what country?
   Bangladesh

10. Snow leopards and tigers live in Jigme Dorji National Park in what country that is seventy-five percent Buddhist?
    Bhutan

11. Peshawar is a major city in what country with the provinces of Sindh and Baluchistan?
    Pakistan

12. K2 is a mountain located in what disputed area?
    Kashmir

13. Harappa is a site in what country containing part of the Indus River?
    Pakistan

14. The Helmand and Amu Darya Rivers are located in what country whose capital is Kabul?
    Afghanistan

15. Nepal is separated from Bangladesh by what corridor in India?
    Siliguri Corridor

16. What is the largest lake in Nepal, declared a Ramsar site in 2007?
Lake Rara

17. What peak in Bhutan has the distinction of being the highest unclimbed peak in the world?
Gangkhar Puensum

18. Bhutan borders what Indian state to the left?
Sikkim

19. Dzongkha is the official language of what country bordering China and India?
Bhutan

20. The lapis lazuli adorning the funeral mask of King Tutankhamen was mined in what mountainous region of Afghanistan?
Hindu Kush

21. Rangpur is a major city in what country known for its common floods?
Bangladesh

22. What country's rail system transports four billion passengers each year across different parts of the country?
India

23. What country, at 330,000 people, is the least populous in South Asia?
Maldives

24. Lakshadweep is a Union Territory west of Kerala, belonging to what country?
India

25. Gujranwala is a city near Sialkot in what country?
Pakistan

26. Chennai is a major city in India, on what bay?
Bay of Bengal

27. Sittwe is a city in what country whose capital is Naypyidaw?
Myanmar

28. What country, along with Thailand and Malaysia, constitutes part of the Malay Peninsula?
Myanmar

29. China claims much of what Indian state's territory?
Arunachal Pradesh

30. The Indo-Gangetic Plain is a region in what part of India – North or South?
North

31. Multan and Faisalabad are major cities in what country bordering the mouth of the Indus River?
Pakistan

32. The Indus River flows through Pakistan, China, and what country?
India

33. Vishakhapatnam is a major Indian city on which coast of India – West or East?
East

34. Ludhiana is the capital of what Indian state known for its majority Sikh population?
Punjab

35. Mon is one of seven states in what country containing the cities of Yangon, Naypyidaw, and Mandalay?
Myanmar

36. Many Dravidian temples are found in Chennai, a seaside city located in the southern part of which Asian country?
India

37. Jamshedpur and Nagpur are cities in what country?
India

38. Palampur is a city in what country?
India

39. The Kabul and Jhelum Rivers flow into what major river rising in southwestern Tibet?
Indus River

40. Himachal Pradesh is part of what mountain range?
Himalayas

41. The Indus River can be found mostly in what country to the southeast of Afghanistan?
Pakistan

42. Taunggyi is a major city in what country?
Myanmar

43. Tiruchchirappalli is a city in what Indian state?
Tamil Nadu

44. Baluchistan is a province in what country that has claim over Kashmir?
Pakistan

45. The Salween River flows through what country whose legislative capital is Yangon?
Myanmar

46. The Khumbu Valley is near Mount Everest in what country?
Nepal

47. The Khyber Pakhtunkhwa is one of the four provinces of what country?
Pakistan

48. The Drangme Chhu River, a tributary of the Manas River, is the lowest point in what country bordering India to the south, east, and west?
Bhutan

49. Bhutan is divided into how many dzongkhags, or districts?
Twelve

50. Wangdue Phodrang is a major town in what country whose largest religions are Buddhism and Hinduism?
Bhutan

51. Addu Kandu, also known as the Addu Channel, is located between Huvadhu Atoll and Addu Atoll in what country?
Maldives

52. The Laccadive Sea borders India, Maldives, and what other country containing the Jaffna Peninsula?
Sri Lanka

53. The Malvathu and Deduru Rivers are in what country primarily inhabited by Sinhalese-speaking and Tamil-speaking people?
Sri Lanka

54. The Punjab Plains, in Pakistan, extend into what country with a state of the same name?
India

55. Chennai, Bengaluru, Hyderabad, Kochi, Coimbatore, Vishakhapatnam, Mysore, and Madurai in no specific order are the eight of the most populous metropolitan areas in what geographical region of India?
South

56. Pulicat Lake is located in Tamil Nadu and what other state home to the Satish Dhawan Space Center?
Andhra Pradesh

57. Tamil is spoken significantly by people in Singapore, Malaysia, India, and what other country containing the Jaffna Peninsula and bordering Palk Strait?
Sri Lanka

58. The Kudahuvadhoo Channel and Veimandu Channel are located in what country south of Lakshadweep?
Maldives

59. The Mouth of the Godavari River is located in what country bordering the Bay of Bengal?
India

60. Royal Bardia National Park is located in what country?
Nepal

61. Lalitpur and Pokhara are major cities in what country, home to Shey Phoksundo National Park?
Nepal

62. The Sundarbans are in India and what country?
Bangladesh

63. Narayanganj and Khulna are major cities in what country?
Bangladesh

64. Ramree Island in Myanmar borders what bay to the north?
Combermere Bay

65. Phrumsengla National Park is located in what landlocked country?
Bhutan

66. Phuntsholing is a major city in what country whose largest religions are Buddhism and Hinduism?
Bhutan

67. Indravati National Park is located in what landlocked Indian state bordering Madhya Pradesh to the west?
Chhattisgarh

68. Gwadar Bay borders Iran and what country?
Pakistan

69. The Sutlej and Ravi Rivers are located in the state of Punjab in what country, home to the cities of Hyderabad and Bahawalnagar?
Pakistan

70. The mouth of the Ayeyarwady River is located in what sea?
Andaman Sea

71. The Miri Hills and Daphla Hills are located in what country?
India

72. The Moscos Islands belong to what country home to Alaungdaw Kathapa National Park?
Myanmar

73. The Gulf of Martaban borders what country to the north, west, and east?
Myanmar

74. The Eastern and Western Ghats border the Deccan Plateau in what country?
India

75. Kathmandu is the capital of what predominantly Hindu nation?
Nepal

76. Dzongkha is the national language of what country bordering Tibet?
Bhutan

77. Mumbai is a city in what Marathi-speaking Indian state?
Maharashtra

78. What Telugu-speaking state is the newest in India and was officially established in 2014?
Telangana

79. Madurai and Chennai are cities in what Indian state bordering Kerala?
Tamil Nadu

80. The Bamiyan Caves are located in what landlocked country bordering Pakistan?
Afghanistan

81. The Karakoram Highway stretches from China to what country?
Pakistan

82. The ancient archaeological site of Mohenjo-Daro is located near the city of Larkana in what country?
Pakistan

83. Multan Cricket Stadium is located in what country whose capital is Islamabad?
Pakistan

84. Pabna and Barisal are cities in what country straddling part of the Ganges River?
Bangladesh

85. The Chittagong Hills are located east of the Karnaphuli Reservoir in what country?
Bangladesh

86. Jigme Dorji National Park, home to Mount Jomolhari, is located in what Himalayan country?
Bhutan

87. The Satpura Range is north of what major Indian plateau?
Deccan Plateau

88. Nanga Parbat is located in the Himalaya Mountains in what country?
Pakistan

89. The Ganges River Plain is south of what country in the Himalayas?
Nepal

90. Kandahar and Herat are major cities in what country containing the Hari River?
Afghanistan

91. The Narmada and Godavari Rivers are located in what country containing part of the Thar Desert?
India

92. The Wakhan Corridor is located in the eastern part within Badakhshan Province in what country?
Afghanistan

93. Andhra Pradesh borders what bay west of Myanmar?
Bay of Bengal

94. Gujarat borders what gulf to the south?
Gulf of Khambhat

95. The Vindhya and Satpura Mountain Ranges are north of the Deccan Plateau and west of what plateau in Eastern India?
Chota Nagpur Plateau

96. Daman and Diu is a Union Territory of what country bordering Pakistan?
India

97. What pass is located at the tripoint of India, Myanmar, and China?
Diphu Pass

98. Nanda Devi is a mountain in the Himalayas in what country?
India

99. The Ravi River is in what country whose major cities include Multan and Peshawar?
Pakistan

100. The Murghab and Khash Rivers are located in what country home to the Paropamisus Mountain Range?
Afghanistan

101. Chilika Lake, the largest lake in India, is located in what Indian state where Oriya is spoken?
Odisha

102. False Divi Point, a headland on the Coromandel Coast, borders the Bay of Bengal in what country?
India

103. The Bhima and Manjra Rivers are located in what country bordering the Gulf of Kutch?
India

104. Imphal is the capital of what Indian state located in the northeast region of the country?
Manipur

105. Valmiki National Park and Wildlife Sanctuary is located in the West Champaran district of what state in India?
Bihar

106. What state, bordering Gujarat and Punjab, is the largest in terms of area in India?
Rajasthan

107. The Open Hand Monument is a symbolic structure in what Indian Union Territory?
Chandigarh

108. Alappuzha is the sixth largest city in what state in southern India?
Kerala

109. Vijayawada is a city on the banks of what major South Indian River partly in Andhra Pradesh?
Krishna River

110. Sanskrit is a philosophical language in Jainism, Buddhism, and what other Indian religion?
Hinduism

111. Konkani is an Indo-Aryan language that is the official language of the state of Goa in what country?
India

112. Kanyakumari, the southern tip of India, is in what state bordering the Bay of Bengal and Kerala?
Tamil Nadu

113. The Central Makran mountain range is located in the southwestern region of what country?
Pakistan

114. Rakhine and Kachin are states in what country home to the Shan Plateau?
Myanmar

115. Sonmiani Bay, south of Pakistan, is northwest of what river delta whose river was the center of the Indus Valley Civilization?
Indus River Delta

116. The Banas and Chambal Rivers are located in what country containing the Coromandel Coast?
India

117. The Ten Degree Channel separates the Andaman Islands from what other island group?
Nicobar Islands

118. The Hingol River empties out into what bay in the Arabian Sea?
Sonmiani Bay

119. Indawygi Lake is located in what country bordering Bangladesh?
Myanmar

120. Nagpur is a city on the edge of what mountain range in India?
Satpura Range

121. Davangere and Shivamogga are cities in what country?
India

122. Dravidian, Sinhalese, and Arab are the main ethnic groups of what archipelagic country bordering the Laccadive Sea to the north?

Maldives

123. Trincomalee is a major resort port city of what province in Sri Lanka?
Eastern Province

124. The Konark Sun Temple is located in what Indian state along the Bay of Bengal?
Odisha

125. Himachal Pradesh is located in what region of India – North or South?
North

126. Carnatic music is from the southern region of what country?
India

127. Kuchipudi is a type of dance originating in Andhra Pradesh in what country?
India

128. Rajkot is a major city in what state bordering Maharashtra?
Gujarat

129. Meghalaya and Nagaland are states in what country?
India

130. Sindh and Punjab are provinces in what country?
Pakistan

131. Minneriya National Park is located in what country bordering the Palk Strait?
Sri Lanka

132. Bengaluru is the capital of what state in India where Kannada is an official language?
Karnataka

133. The Sundarbans are shared by Bangladesh and what other country?
India

134. Port Blair is a city on what island in the Andaman and Nicobar Islands?
South Andaman

135. What city is the capital of Tamil Nadu, India's fourth largest city by population, and the most populous city in South India?
Chennai

136. Indravati National Park is located in what state bordering Jharkhand to the northeast?
Chhattisgarh

137. Shillong is the capital of what Indian state in the northeast?
Meghalaya

138. Vadodara and Surat are cities in what state bordering the Gulf of Kutch?
Gujarat

139. Vishakhapatnam is the most populous city in what South Indian state?
Andhra Pradesh

140. The Palamau Tiger Reserve and Betla National Park are located in what Indian state?
Jharkhand

141. The Balaghat Mountain Range is located in what state bordering Karnataka?
Maharashtra

142. Mizoram borders Bangladesh and what other country?
Myanmar

143. Kanha National Park is located in the eastern region of what Indian state?
Madhya Pradesh

144. Allahabad and Kanpur are major cities in what state in North India?
Uttar Pradesh

145. Hemis National Park is located in what Indian state?
Jammu and Kashmir

146. Amritsar and Ludhiana are the most populous cities in what state?
Punjab

147. Bhakra Dam is located in what state whose capital is Shimla?

Himachal Pradesh

148. Maduru Oya National Park is located in what country bordering Koddiyar Bay?
Sri Lanka

149. Hingol National Park is located in what province in Pakistan?
Baluchistan

150. Kirthar National Park is located in what Pakistani province?
Sindh

151. Royal Bardia National Park is located in what landlocked country?
Nepal

152. Haora is a major city in what Indian state whose capital is Kolkata?
West Bengal

153. Narayanganj is a major city south of Dhaka in what country?
Bangladesh

154. What state in India is the only one where Sikhs form the majority of the population?
Punjab

155. Pench National Park is located in what state whose capital is Bhopal?
Madhya Pradesh

156. Bhubaneshwar is the capital of what state in India bordering the Bay of Bengal and Chhattisgarh?
Odisha

157. People speak Telugu primarily in Telangana and what other South Indian state?
Andhra Pradesh

158. Bengaluru and Mysore are the two largest cities in what state bordering the Arabian Sea?
Karnataka

159. Sholapur, Pune, Thane, Aurangabad, Nagpur, and Kalyan are six of the most populous cities in what Indian state whose capital and most populous city is Mumbai?
Maharashtra

160. Chennai, the fourth most populous city in India, is the most populous in South India and in what state whose coast is on the Bay of Bengal?
Tamil Nadu

161. The Brihadeeswarar Temple is a Hindu Temple located in the city of Thanjavur in Tamil Nadu in what country?
India

162. The Connaught Palace is the largest financial, commercial, and business center in India, and is located in what major city?
Delhi

163. The Akshardham Temple in New Delhi, located near the banks of the Yamuna River, is located in what country?
India

164. The Qutb Minar, the tallest brick minaret in the world, is a UNESCO World Heritage Site in what city in India?
New Delhi

165. The Taj Mahal Palace Hotel is located next to the Gateway of India in what city in the state of Maharashtra in India?
Mumbai

166. Arma Konda is the highest point in what mountain range west of the Coromandel Coast in India?
Eastern Ghats

167. Patna, Varanasi, and Allahabad are major cities on what long river in India in the northern part of the country?
Ganges River

168. The Godavari River, whose source is at Brahmagiri Mountain near Trimbakeshwar Shiva Temple, has its mouth in what body of water that is part of the Indian Ocean?
Bay of Bengal

169. The Palk Strait is located between the Mannar District of the Northern Province of Sri Lanka and what state in India?
Tamil Nadu

170. The Kapaleeshwarar Temple is a Hindu Temple of Shiva located in Mylapore in what major city in Tamil Nadu that is the fourth largest metropolitan area in India

Chennai

171. The Bumdeling Wildlife Sanctuary covers most of the Trashiyangtse District in what country whose capital is Thimphu?
Bhutan

172. Bangladesh is separated from Nepal and Bhutan by what corridor located in India?
Siliguri Corridor

173. The Shwenandaw Monastery is a historic Buddhist monastery located in Mandalay in what country?
Myanmar

174. Bala Hissar, a fortress in Kabul, is situated on Kuh-e-Sherdarwaza Mountain in what country?
Afghanistan

175. Kyichu Lhakhang is an important Buddhist temple in the Paro District of what landlocked South Asian country?
Bhutan

176. The Chittagong Hills are located in the southeastern region of what small South Asian country?
Bangladesh

177. Paro Taktsang is a Himalayan Buddhist Site in what valley in Bhutan?
Paro Valley

178. The Surma River, part of the Surma-Meghna River System, is a major river in what South Asian country, one of the most densely populated countries in the world?
Bangladesh

179. What major city in Bangladesh was the 2012 ISESCO Asian Capital of Culture?
Dhaka

180. The Pyu City-States, some of which are now UNESCO World Heritage Sites, are located in the central part of what country?
Myanmar

181. The Jatiyo Sangsad Bhaban is the national parliament house of what country whose capital is Dhaka?
Bangladesh

182. The Tamzhing Monastery is located in the Bumthuma District in the central region of what landlocked South Asian county?
Bhutan

183. The Changlimithang Stadium is located in Thimphu and is the home of the National Football Team in what country?
Bhutan

184. A gemstone called Lapis lazuli has been mined in the Kokcha River Valley for thousands of years in what country?
Afghanistan

185. Shahjalal International Airport is located in the major city of Dhaka in what South Asian country?
Bangladesh

186. The Shwedagon Pagoda is a famous Buddhist temple located in Yangon in what South Asian country?
Myanmar

187. Patenga is a beach located near the mouth of the Karnaphuli River near Chittagong in what country?
Bangladesh

188. What Bangladeshi city, located on the banks of the Shitalakshya River, is a major city in this coastal country and is nicknamed the "Dundee of Bangladesh"?
Narayanganj

189. What temple located in the village of Minnanthu in Myanmar is one of the most visited in Bagan?
Sulamani Temple

190. Shapla Square is located in the Motijheel Thana of what country bordering the Bay of Bengal to the south and whose capital is Dhaka?
Bangladesh

191. What major city, home to the Rose Garden Palace in Bangladesh, is known as the "Rickshaw Capital of the World"?
Bangladesh

192. The Kyaiktiyo Pagoda is a pilgrimage site for Buddhists in the Mon State of what country whose capital is Naypyidaw?
Myanmar

193. The Somapura Mahavihara is a UNESCO World Heritage Site and a Buddhist vihara in what country in South Asia?
Bangladesh

194. The Lalbagh Fort is an incomplete fort complex on the banks of the Buriganga River in the southwestern part of what city in Bangladesh?
Dhaka

195. Hamid Karzai International Airport is located in what city in the Kabul province of Afghanistan?
Kabul

196. The Sixty Dome Mosque is a famous and large mosque that is a UNESCO World Heritage Site, located in the city of Bagerhat in what country?
Bangladesh

197. The Dacht-i-Navar Group is a volcanic massif southwest of Kabul, the capital of what South Asian country?
Afghanistan

198. The Ananda Temple is located in the ancient city of Bagan in what present day country?
Myanmar

199. Lawachara National Park is a major national park in what country that is the eighth most populous in the world?
Bangladesh

200. The Dhammayangyi Temple, the largest in the ancient city of Bagan, was built during the reign of King Narathu in what country?
Myanmar

201. Madhabkunda Falls is a major waterfall in Sylhet in what country in South Asia bordering the Bay of Bengal, Myanmar, and India?
Bangladesh

202. The city of Rajshahi, located on the north bank of the Padma River, is situated close to what country's border with India?
Bangladesh

203. What major city in western Bangladesh is known as the "Silk City", and is the site of the Shah Makhdum Airport?
Rajshahi

204. The Botataung Pagoda, located near the Yangon River, is in the city of Yangon in what country?
Myanmar

205. The Mosque City of Bagarhat is a UNESCO World Heritage Site, situated in the Khulna Division of what South Asian country?
Bangladesh

206. The Ahsan Manzil is a magnificent building located along the banks of the Buriganga River in what city in Bangladesh?
Dhaka

207. The Friday Mosque is located in what major city in Afghanistan that is situated on the Hari River?
Herat

208. The Dhakeshwari Temple is a Hindu Temple located in what city in the South Asian country of Bangladesh?
Dhaka

209. The Lingkana Palace is the home of the king and queen of what country whose capital is Thimphu?
Bhutan

210. The Bangladesh National Museum, which was established and inaugurated in 1913, is located in what city in Bangladesh?
Dhaka

211. The Mahamuni Buddha Temple is situated southwest of Mandalay, the second most populous city in what country?
Myanmar

212. Bhutan borders what state in northeastern India to the west?
Sikkim

213. Sonargaon, formerly the administrative center of the region of Bengal, is located in the center of the Ganges Delta near the city of Narayanganj in what country?
Bangladesh

214. The ruins of Drukgyal Dzong, previously a Buddhist monastery and fortress, is located in the Paro District in what country?
Bhutan

215. Lake Hamun is located in the Registan Desert, in what basin in southwestern Afghanistan?
Sistan Basin

216. The confluence of the Mo Chhu and Po Chhu Rivers is at the Punakha Dzong in what country?
Bhutan

217. The Star Mosque is located in the Armanitola area of what major city in Bangladesh that is the largest in the country by population?
Dhaka

218. What Buddhist pagoda in Mandalay lies at the foot of Mandalay Hill and is home to the world's largest book?
Kuthodaw Pagoda

219. The Kakrail Mosque is located near the Ramna Park in the major city of Dhaka in what South Asian country?
Bangladesh

220. The Haa Valley is located in the Haa District in the western part of what landlocked South Asian country?
Bhutan

221. The Kantajew Temple is a medieval Hindu Temple in the city of Dinajpur in what South Asian country bordering the Bay of Bengal?
Bangladesh

222. The Rinpung Dzong is a large Buddhist monastery located in the Paro District of what country?
Bhutan

223. The Sher-e-Bangla Stadium, which is also known as the Mirpur Stadium, is a cricket stadium located in what major city in Bangladesh?
Dhaka

224. Tashichho Dzong is a Buddhist monastery on the Wang Chu River in what country whose capital is Thimphu?
Bhutan

225. Saka Haphong is one of the highest peaks in what country bordering Myanmar and India?
Bangladesh

226. The Bangabandhu Bridge, also known as the Jamuna Multipurpose Bridge, is located across the Yamuna River and connects Bhuapur to Sirajganj in what South Asian country?
Bangladesh

227. The Black Mountains are located in the central region of what country that is home to Jigme Singye Wangchuck National Park?
Bhutan

Southeast Asia
Thailand, Laos, Cambodia, Vietnam, Malaysia, Singapore, Brunei, Philippines, Indonesia, East Timor

1. The Gulf of Thailand borders what country to the north and west?
Thailand

2. The Annamese Cordillera forms the border between Laos and what other country?
Vietnam

3. Ho Chi Minh City is the largest city in what country?
Vietnam

4. Tonle Sap is a lake in what country bordering Thailand?
Cambodia

5. Suoi Tien Cultural Theme Park, a water park devoted to Buddhism, can be found in what country?
Vietnam

6. The Petronas Twin Towers, among the tallest buildings in the world, are located in what Malaysian city?
Kuala Lumpur

7. Sentosa is a popular tourist attraction in what country south of Malaysia?
Singapore

8. A Hindu statue stands outside the Batu Caves in Selangor in what country?
Malaysia

9. Taman Negara National Park is the largest national park in what country containing the Pahang River?
Malaysia

10. Bandar Seri Begawan is the capital of what country?
Brunei

11. Underwater World and Dolphin Lagoon are attractions in what country?
Singapore

12. Chinatown Heritage Center is located in what capital city?
Singapore

13. Nino Konis Santana National Park is located in what country?
East Timor (Timor-Leste is acceptable)

14. Denpasar is the capital of what Hindu island in Indonesia?

Bali

15. Puncak Mandala and Jaya Peak are peaks on the western part of the island of New Guinea in what country?
Indonesia

16. Sulawesi is an island in what country bordering the Makassar Strait?
Indonesia

17. Dili is the capital of what country?
East Timor (Timor-Leste is acceptable)

18. Luagan Merimbun is a site in what country?
Brunei

19. Mount Mayon, which has erupted 47 times in the past 400 years, is located in Luzon in what country?
Philippines

20. Davao is a major city in what country bordering the Sulu Sea to the west and south?
Philippines

21. The Natuna Sea borders what country whose capital is Jakarta?
Indonesia

22. The Malay Peninsula is situated on what major shelf?
Sunda Shelf

23. Tanjung Piai is a cape in Johor in what country bordering Thailand?
Malaysia

24. The Gulf of Boni feeds into what sea bordering the island of Sulawesi in Indonesia?
Banda Sea

25. The Doberai Peninsula is bordered by the Ceram Sea in what country?
Indonesia

26. Flamingo Bay borders what major island in Indonesia bordering Cenderawasih Bay and Yos Sudarso Bay?
New Guinea

27. Cape Bojeador is located on what island in the Philippines bordering Divilacan Bay?
Luzon

28. The Indonesian islands of Buru and Ceram border what sea to the south?
Banda Sea

29. The Bight of Bangkok is an inlet of the Gulf of Thailand. This bight borders what country?
Thailand

30. Virachey National Park is located in what country whose capital is Phnom Penh?
Cambodia

31. The Bolaven Plateau and Xiangkhoang Plateau are located in what landlocked country?
Laos

32. The Phraya Lowlands can be found in what country home to Khao Yai National Park?
Thailand

33. Thale Luang is a lake in what country bordering Malaysia?
Thailand

34. George Town is a city on the island of Pinang in what country?
Malaysia

35. The Luang Prabang Range in Thailand extends into what country?
Laos

36. The Isthmus of Kra is in what country bordering Malaysia?
Thailand

37. Ko Chang is an island belonging to what country to the north?
Thailand

38. Bandar Seri Begawan is the capital of what small country north of Malaysia?
Brunei

39. Taman Negara National Park is located in what country bordering the Balabac Strait and Brunei Bay?

Malaysia

40. Tonle Sap is a lake in what country bordering Thailand and Vietnam?
Cambodia

41. Ceram and Sulawesi are islands located in what country bordering the Java Sea?
Indonesia

42. Sarawak is located in what country bordering Thailand and Myanmar?
Malaysia

43. The Gulf of Tonkin borders what country to the west?
Vietnam

44. Luang Prabang is a province in what country whose highest point is Phou Bia?
Laos

45. The Straits of Johor separates Malaysia from what country home to the islands of Pulau Tekong and Pulau Ayer Chawan?
Singapore

46. West Nusa Tenggara is a province of what archipelagic country?
Indonesia

47. Moro is a gulf forming part of the southern coast of Mindanao in what country that gained independence on July 4, 1946 from the United States?
Philippines

48. Phou Bia is a peak located in Laos on what major plateau bordering the Annamite Range?
Xiangkhoang Plateau

49. Samar is an island off the southeastern coast of what country?
Philippines

50. Davao is the most populous city on what island in the Philippines?
Mindanao

51. Cu Chi is a site in what country containing the Hong and Da Rivers?
Vietnam

52. Palawan borders the Sulu Sea to the east and the South China Sea to the west. This island belongs to what archipelagic country?
Philippines

53. The fertile floodplain of the Chao Phraya River is the chief rice-growing region in what country?
Thailand

54. Sulawesi, the world's eleventh-largest island, is located in what country?

Indonesia

55. Cape Ron is in what country bordering the Gulf of Tonkin and the South China Sea?
Vietnam

56. Yak Loum is a lake and a popular tourist destination in the Ratanakiri province of what country?
Cambodia

57. The Visayan Sea borders what island to the north containing the city of Masbate?
Masbate

58. Kinabalu is the highest point in what country?
Malaysia

59. The Minahasa Peninsula borders what gulf that feeds into the Molucca Sea?
Gulf of Tomimi

60. The Chiang Mai Zoo is located in what country bordering the Gulf of Thailand?
Thailand

61. Damnoen Saduak is a site in the central region of what country?
Thailand

62. Nino Konis Santana National Park is located in what country bordering the Timor Sea?
East Timor

63. The Barat Daya Islands belong to what country whose capital is Jakarta?
Indonesia

64. What island is the most populous in Indonesia?
Java

65. The Makassar Strait separates Borneo from what large island?
Sulawesi (Celebes is acceptable)

66. The Central Highlands are located in what country bordering the South China Sea?
Vietnam

67. The Pahang and Rajang Rivers are located in what country whose capital is Kuala Lumpur?
Malaysia

68. The Batu Caves are located in Selangor in what country?
Malaysia

69. Tasek Merimbun is the largest lake in what country?
Brunei

70. Cape Datu is located in what country bordering the Natuna Sea and the South China Sea?
Malaysia

71. The Luang Prabang Mountain Range can be found in the western region of what country?

Laos

72. Tioman and Banggi are islands belonging to what country?
Malaysia

73. Nam Ngum Dam is located in what country?
Laos

74. Dao Phu Quoc is an island belonging to what country?
Vietnam

75. Labuk Bay is located in the Sulu Sea east of what country?
Malaysia

76. The Mekong River Delta is located in what country?
Vietnam

77. The Cagayan Islands and Quiniluban Islands are located in what country bordering Leyte Gulf and the Bohol Sea?
Philippines

78. Siem Reap and Batambang are major cities in what country?
Cambodia

79. The Cardamom Mountains are located in what country?
Cambodia

80. The Petronas Twin Towers, among the tallest in the world, are located in Kuala Lumpur, the capital of what country where Islam, Hinduism, Buddhism, and Christianity are the four largest religions?

A Competitor's Compendium to the Geography Bee

Malaysia

81. Brunei is completely surrounded by what Malaysian State?
Sarawak

82. The district of Limbang in the Malaysian State of Sarawak separates the two parts of what country?
Brunei

83. The Temburong District is a mountainous part of what country whose capital is Bandar Seri Begawan?
Brunei

84. Brunei International Airport, located in Brunei, is situated in what major city?
Bandar Seri Begawan

85. Angkor Wat, a temple complex in Southeast Asia, is the largest religious monument in the world and is located in what country?
Cambodia

86. Bayon is a Khmer temple in the site of Angkor in what Southeast Asian country bordering the Gulf of Thailand?
Cambodia

87. Preah Monivong National Park is home to the Bokor Hill Station in what country?
Cambodia

88. The Damrei Mountains, also known as the Elephant Mountains, is an extension of the Cardamom Mountains in what country?

Cambodia

89. The Sultan Omar Ali Saifuddin Mosque is located in Bandar Seri Begawan, the capital of what Southeast Asian country?
Brunei

90. Botum Sakor National Park is located on a peninsula south of the Cardamom Mountains in what Southeast Asian country?
Cambodia

91. Mount Ramelau is a peak in what country whose two most populous cities are Dili and Same?
East Timor

92. The town of Senmonorom is the capital of the Mondulkiri Province of what country in Southeast Asia?
Cambodia

93. The Dangrek Mountains form the natural border between Cambodia and what other country bordering the Gulf of Thailand?
Thailand

94. Botum Sakor National Park, the largest national park by area in Cambodia, is situated on what gulf?
Gulf of Thailand

95. Istana Nurul Iman, located in Bandar Seri Begawan, is a palace and the official residence of the Sultan of what country?
Brunei

96. The Cablac Mountains are located in what country bordering the Banda Sea?
East Timor

97. Jerudong Park is a famous amusement park in what country whose capital is Bandar Seri Begawan?
Brunei

98. What country, located in the Indomalayan ecozone is home to Tonle Sap, a major lake in Southeast Asia?
Cambodia

99. Phnom Aural, a mountain in the eastern Cardamom Mountains, is the highest peak in what country?
Cambodia

100. The exclave of Ambeno, which borders Indonesia, belongs to what Southeast Asian country?
East Timor

101. The Malay Technology Museum is located in what major city in the Brunei-Muara District in Brunei?
Bandar Seri Begawan

102. The Karimata Strait, which separates the islands of Borneo and Sumatra, connects the South China Sea to what other sea?
Java Sea

103. Kirirom National Park, located in Kampong Speu Province and Koh Kong Province, is in what country?
Cambodia

104. The Kapuas River, located on the island of Borneo, is the longest river in Indonesia and flows through what city that is the capital of the West Kalimantan Province?
Pontianak

105. Mount Tambora is the highest point on what island in the West Nusa Tenggara Province that is part of the Lesser Sunda Islands?
Sumbawa Island

106. Sultan Hassanal Bolkiah Stadium, used mostly for football matches, is located in Bandar Seri Begawan, the capital of what country that borders the South China Sea?
Brunei

107. Koh Rong Sanloem is an island off the coast of Sihanoukville, a city in what country bordering the Gulf of Thailand to the southwest?
Cambodia

108. Siem Reap International Airport, situated near Angkor Wat, is located in the city of Siem Reap and is the busiest airport in what country?
Cambodia

109. Stung Teng, the capital of the Stung Teng Province, is located in Virachey National Park in what country?
Cambodia

110. Atauro Island, situated in the southern part of the Banda Sea, belongs to what country occupying part of the island of Timor?
East Timor

111. Samarinda, which lies on the banks of the Mahakam River, is the most populous city in the East Kalimantan Province of what archipelagic country in Southeast Asia?
Indonesia

112. The Wat Phnom is a Buddhist temple or pagoda located in what major city in Cambodia?
Phnom Penh

113. What country, known as the Emerald of the Equator, is home to the city of Surabaya, which is located on the Kali Jagir River in the East Java Province?
Indonesia

114. The Billionth Barrel Monument is located in Seria, a town in the Belait District of what country bordering Malaysia?
Brunei

115. Sambor Prei Kuk is a site in the Kampong Thom Province of what country with the cities of Sihanoukville and Battambang?
Cambodia

116. The island of Samosir is located in what volcanic lake on the island of Sumatra, in Indonesia?
Lake Toba

117. The Phnom Aural Wildlife Sanctuary in Cambodia is located in what mountain range?
Cardamom Mountains

118. The Bassac River, which begins in Phnom Penh in Cambodia, is a distributary of Tonle Sap and what river?

Mekong River

119. The Monivong Bridge is located on the Bassac River in what major city in Cambodia?
Phnom Penh

120. Ulu Temburong National Park, located in the Temburong District, is what country's first established national park?
Brunei

121. The Royal Palace is located in the major city of Phnom Penh in what country?
Cambodia

122. The Silver Pagoda is a Buddhist temple in the city of Phnom Penh in what country situated next to the Gulf of Thailand?
Cambodia

123. The exclave of Ambeno in East Timor borders what sea to the north which is situated next to the islands of Alor, Pantar, and Lombien?
Savu Sea

124. The sources of the Hari River and Musi River lie in the Barisan Mountains, a mountain range located in the southwestern part of what island in Indonesia?
Sumatra

125. Battambang and Kampong Tham are cities in what Southeast Asian country bordering Laos, Vietnam, and Thailand?
Cambodia

126. Koh Ker is a remote site in the northern part of what country in Southeast Asia whose capital and largest city is Phnom Penh?
Cambodia

127. Phnom Kulen National Park is located in the Siem Reap Province in what country bordering Laos and the Gulf of Thailand?
Cambodia

128. The Wetar Strait separates East Timor and the island of Timor from what island in Indonesia?
Wetar

129. Banteay Srei is a 10th century Hindu Temple dedicated to the Hindu god Shiva in what present day country?
Cambodia

130. Patenggang Lake is located in what city in western Java that is the third most populous city in Indonesia?
Bandung

131. Brunei Bay is located on the northwestern coast of what island on which Brunei is located?
Borneo

132. The Cambodian Cultural Village is a museum and a theme park in what major city in Cambodia?
Siem Reap

133. Lake Ira Lalaro is the largest freshwater lake in what country bordering Indonesia?
East Timor

134. Koh Thonsay, an island in the Gulf of Thailand, translates to Rabbit Island and belongs to what Cambodian province?
Kep Province

135. The Mount Paitchau Important Bird Area is located in what country whose highest point is Mount Ramelau?
East Timor

136. Wat Phnom is a famous Buddhist temple located in the city of Phnom Penh, the capital of what Southeast Asian country?
Cambodia

137. What island in East Timor is East Timor's easternmost point, and is a major tourist attraction in East Timor?
Jaco Island

138. The Raja Ampat Islands, located off the northwestern coast of the island of New Guinea and the Bird's Head Peninsula in what country whose capital is Jakarta?
Indonesia

139. Chaktomouk Hall, located on the banks of the Tonle Sap, is located in what major city in Cambodia?
Phnom Penh

140. What country, located on the island of Borneo, and bordering the South China Sea, hosted the 1999 Southeast Asian Games?
Brunei

141. Wat Ounalom, located near the Royal Palace of Cambodia and the center of Buddhism in Cambodia, is located in what major city?
Phnom Penh

142. The Sorya Shopping Center is a large shopping mall in Phnom Penh, the capital of what Southeast Asian country bordering Vietnam to the east?
Cambodia

143. Nino Konis Santana National Park is the first national park established in what country whose capital is Dili?
East Timor

144. The Meratus Mountains, which almost divides the South Kalimantan Province into two equal parts and is located on what Indonesian island?
Borneo

145. The Preah Vihear Temple is an ancient Hindu Temple and a UNESCO World Heritage Site in what country?
Cambodia

East Asia
China, Japan, North Korea, South Korea, Taiwan, Mongolia

1. Hallasan is the highest point in what country bordering the Jeju Strait?
South Korea

2. Tibetan Buddhism and Lamaism are followed in what nomadic country?
   Mongolia

3. The Taklimakan Desert is located in the Xinjiang Uyghur Autonomous Region. This administrative division belongs to what country home to the Dzungarian Basin?
   China

4. Toyama Bay forms part of the northern coast of what country?
   Japan

5. The Greater Khingan Range is situated west of the Gan and Nen Rivers in what country bordering Hangzhou Bay?
   China

6. The Lesser Khingan Range in China is northwest of what major plain?
   Manchurian Plain

7. Kaohsiung is a major city on what island north of the Luzon Strait?
   Taiwan

8. The Sichuan Basin is west of Chongqing in what country?
   China

9. Pyongyang is the capital of what country?
   North Korea

10. The Korea Strait separates South Korea from what archipelagic country?
Japan

11. The La Perouse Strait separates Sakhalin from what large Japanese island?
Hokkaido

12. Incheon is one of the largest cities in what country bordering North Korea?
South Korea

13. Korea Bay borders China and what country?
North Korea

14. What city, located in the northern region of Taiwan, is the capital of this sovereign state?
Taipei

15. Yushan National Park is located in what sovereign state?
Taiwan

16. Guiyang is a city in what Chinese province bordering Guangxi Zhuang and Jiangxi?
Hunan

17. Busan, Gwangju, and Daejeon are major cities in what country bordering the Yellow Sea?
South Korea

18. Daisen-Oki National Park is located in what country bordering the Sea of Japan?
Japan

19. The Qiongzhou Strait separates what island from mainland China?
Hainan

20. Handong and Hebei are provinces in what country bordering the Gulf of Tonkin?
China

21. What city is the capital of the Chinese province of Sichuan?
Chengdu

22. Fuzhou is the capital of what province bordering the Taiwan Strait to the southeast?
Fujian

23. What city in South Korea is the country's largest city after Seoul and Busan?
Incheon

24. Tsushima Island is in the Korea Strait and belongs to what country to the east?
Japan

25. Hohhot is the capital of what autonomous region in China?
Inner Mongolia

26. Qinghai Lake is located in what country bordering Bo Gulf?
China

27. The Gobi Desert and Altay Mountains are located in what landlocked country?
Mongolia

28. The Kuntun Mountains are north of what plateau in China?
Plateau of Tibet

29. Dongting Lake is located in what country with the Great Khingan Range?
China

30. Hainan belongs to what country straddling the Yunnan Plateau?
China

31. The Sichuan Basin is located in what country home to the Daxue Mountains?
China

32. The Eastern Sayans Mountain Range extends from Russia into what country whose highest point is Nayramadlin Orgil?
Mongolia

33. The Nei Mongol Plateau in China and Mongolia is a desert plateau north of what major river that was a part of Ancient China?

Yellow River

34. The Karimata Strait connects the Java Sea to what other sea?
South China Sea

35. What peak is the highest in the Qin Mountains at 3,767 meters?
Mount Taibai

36. The Wei River is the largest tributary of what river in China?
Yellow River

37. The Bayan Har Shan and Qilian Shan are mountain ranges southeast and northeast of Qaidam Basin in what country?
China

38. Songnisan National Park is located west of Sobaek Sanmaek in what country?
South Korea

39. Hiroshima Peace Memorial Park is located in what country bordering the Inland Sea?
Japan

40. What city in China is located at the confluence of the Jialing and Yangtze Rivers?
Chongqing

41. Heaven Lake, part of the Baekdudaegan mountain range, is located on China's border with what country?
North Korea

42. The Hengduan Shan in northern Myanmar extend into what country?
China

43. The Shinano River is the longest river in what archipelagic country?
Japan

44. The Tarim Basin is south of the Tian Shan in what country?
China

45. The Mu Us Desert in China is south of the Yin Shan and what river?
Yellow River

46. Leizhou Bay borders what country to the west and north?
China

47. In which Asian country would you find kabuki actors using music and colorful, elaborate costumes to tell a story?
Japan

48. Cho Oyu is a mountain on Nepal's border with what country?
China

49. Tsugaru is a strait forming part of the northern coast of what country?
Japan

50. East Korea Bay, on the eastern coast of North Korea, feeds into what sea?
Sea of Japan

51. Khan Tengri is the highest peak in Kazakhstan on the country's border with China. What mountain range is this peak located in?
Tian Shan Mountains

52. Hustai National Park is located in what country home to the Hangayn Mountains and the Herlen River?
Mongolia

53. Sendai is a major city on what large Japanese island?
Honshu

54. Lhasa is a city in what Chinese province bordering Qinghai and Sichuan?
Tibet

55. The Altun Shan Mountains are located east of the Taklimakan Desert in what country?
China

56. The yuan is the national currency of what country?
China

57. Kushiro Shitsugen National Park is located on the island of Hokkaido in what country?
Japan

58. The Izu Islands belong to what country?
Japan

59. Kochi is a city on the island of Shikoku in what country?
Japan

60. The Nemegt Basin is in the southern region of what country bordering Russia to the north and China to the south?
Mongolia

61. Ulsan and Busan are major cities in what country?
South Korea

62. The Changlin Reservoir is located in what country?
North Korea

63. Tsonjin Boldog is located in what landlocked country?
Mongolia

64. The Oki Islands are located in what country?
Japan

65. The Gobi Desert is shared by China and what other country?
Mongolia

66. China claims land primarily in what state in northeastern India?
Arunachal Pradesh

67. The Greater Khingan Mountain Range is located in what country bordering the South China Sea?
China

68. Uvs Nuur and Har Nuur are lakes in what country?
Mongolia

69. The Sup'ung Reservoir is shared by China and what other country?
North Korea

70. The Korea Strait separates what two countries?
Japan and South Korea

71. Songni Mountain National Park and the Gatbawi Statue are located in what country?
South Korea

72. The Three Gorges Dam, located on the Yangtze River, is located in what country?
China

73. The Bayan Har Shan is a mountain range in what country?
China

74. Taichung is a city on what island?
Taiwan

75. The Tsushima Strait separates the island of Tsushima from what country bordering Uchiura Bay?
Japan

76. Saitama and Kawasaki are major cities on Honshu in what country?
Japan

77. The Osumi Islands and Tokara Islands belong to what country?
Japan

78. The Bungo Strait separates Shikoku from what major island?
Kyushu

79. The Qarqan and Tarim rIvers are located in what country bordering the East China Sea?
China

80. Lanzhou, Zhengzhou, and Shijiazhuang are major cities in what country bordering Mongolia, Russia, India, North Korea, and Vietnam?
China

81. The Bohai Sea, a part of the Yellow Sea, borders what country to the west?
China

82. The Terracotta Warriors site is located in the Shaanxi Province in what country?

China

83. The Longsheng Rice Terraces are located in the Guangxi Province in what country bordering the South China Sea?
China

84. The Li River flows from the city of Guilin to the city of Yangshou in what country?
China

85. The Changbai Mountains separate China from what country to the east?
North Korea

86. The Taihang Mountains in China are located on the eastern edge of what plateau also known as the Huangtu Plateau?
Loess Plateau

87. The Gurbantunggut Desert is located in the northern part of the Xinjiang Uyghur Autonomous Region in what major basin?
Dzungarian Basin

88. Heaven Lake, a crater lake on the border between China and North Korea, lies within a caldera on what mountain?
Paektu Mountain

89. The Qin Mountains, which divide North and South China, are located south of what densely populated river valley?
Wei River Valley

A Competitor's Compendium to the Geography Bee

90. The Wuyi Mountains are located in the Nanping Prefecture of what Chinese Province?
Fujian Province

91. The Mogao Caves are a UNESCO World Heritage Site in what province in China?
Gansu Province

92. Lushan National Park is located west of Poyang Lake in the Jiangxi Province in what country?
China

93. The Longmen Grottoes, also known as the Longmen Caves, is home to more than sixty Buddhist pagodas in what East Asian country?
China

94. The Summer Palace is home to Kunming Lake and Longevity Hill, and is a major tourist destination in Beijing in what country?
China

95. Tokyo is located on the banks of what river in Japan that flows into Tokyo Bay?
Sumida River

96. What city in Japan, known as the "City of 10,000 Shrines" was formerly the capital of Imperial Japan for more than one thousand years?
Kyoto

97. Sapporo, a major city located on the island of Hokkaido in Japan, hosted the 1972 Winter Olympics and is located on what plain?

Ishikari Plain

98. Fukuoka, the most populous city on the island of Kyushu, is located on what sea?
Sea of Genkai

99. The Matsuyama Castle is a major point of interest in what city on the island of Shikoku that is famous for its hot springs?
Matsuyama

100. The Toyota car company began in what city located near the Kiso River in the Aishi Prefecture of Japan, on the island of Honshu?
Nagoya

101. The Jigokudani Valley is located in Joshinetsu Kogen National Park on the island of Hokkaido in what country?
Japan

102. The Historic Monuments of Ancient Kyoto are a UNESCO World Heritage Site in what East Asian country?
Japan

103. Tokyo, which will host the 2020 Winter Olympics, is located in the Tokyo Prefecture and is the capital of what East Asian country?
Japan

104. The metropolitan area of Osaka-Kobe consists of two cities located in two different prefectures in the Kansai Region of what country?
Japan

Central Asia and Asian Russia
Kazakhstan, Uzbekistan, Tajikistan, Kyrgyzstan, Turkmenistan, Asian Russia

1. Lake Balkhash is located in what country?
   Kazakhstan

2. Turkmenabat is a city in what country bordering Garabogaz Bay?
   Turkmenistan

3. Lake Ysyk is located in what country?
   Kyrgyzstan

4. Lake Tengiz is in the north-central region of what country?
   Kazakhstan

5. Kugitang Nature Reserve, home to 438 dinosaur footprints, can be found in what country?
   Turkmenistan

6. Shahristan is an ancient town in what country bordering Kyrgyzstan and Uzbekistan?
   Tajikistan

7. The Singing Dunes can be found in Altyn-Emel State National Park in what country?
   Kazakhstan

8. Ashgabat is the capital of what country?
   Turkmenistan

9. The Yangiabad Rocks can be found in Tashkent, in what country?
   Uzbekistan

10. What is the capital of Kyrgyzstan?
    Bishkek

11. Garabogaz Bay is an inlet of what lake?
    Caspian Sea

12. The Syr Darya is a river in what country that is home to the Khan Shatyr tent?
    Kazakhstan

13. Chelyabinsk is a major city miles south of Yekaterinburg in what country?
    Russia

14. The Yenisey and Kheta rivers can be found in what country?
    Russia

15. The world's largest open-pit gold mine is at Muruntau in what country?
    Uzbekistan

16. The Baikonur Cosmodrome, site of most space flights launched by the Soviet Union, is located in what country?
    Kazakhstan

17. More than 90% of Tajikistan is covered by the Tian Shan and what other mountain range?
    Pamirs

18. The Baikonur Cosmodrome is managed by what agency?
    Russian Federal Space Agency

19. Uzbekistan lies on what famous road that was a major point of trade in ancient times?
    Silk Road

20. The Caspian Depression is found in Russia and what country?
    Kazakhstan

21. The Ustyurt Plateau is a basin in what two countries?
    Kazakhstan and Uzbekistan

22. The Kazakh Uplands are located in what country?
    Kazakhstan

23. The Alay Valley is located in what country bordering Uzbekistan to the west?
    Kyrgyzstan

24. The Torghay River is in what country bordering the Caspian Sea?
    Kazakhstan

25. Ergaki National Park is located in what country bordering Shelikhov Gulf?

Russia

26. Lake Andreyevskoye is located in what country containing the Verkhoyansk and Chersky Mountains?
Russia

27. The Mir Mine is located in what country bordering Khatanga Gulf and the Gulf of Ob?
Russia

28. Balkanabat is the capital of Balkan Province in what country bordering the Caspian Sea to the west?
Turkmenistan

29. The Alay Mountains and Zarafshan Range are located in what country home to the Kulma Pass?
Tajikistan

30. The Toktogul Reservoir is located in the western region of what country home to Ala Archa National Park?
Kyrgyzstan

31. The Wudang Shan are located in China, which borders what country whose capital is Bishkek?
Kyrgyzstan

32. Altun Emel National Park is located in Almaty in what country?
Kazakhstan

33. Lake Tengiz is miles southwest of the Esil River in what country bordering Uzbekistan?

Kazakhstan

34. The Yangiabad Rocks are located in what country with the cities of Samarkand and Bukhara?
Uzbekistan

35. The Turan Lowland extends across three countries in Central Asia. Name the countries.
Kazakhstan, Uzbekistan, Turkmenistan

36. The Sikhote-Alin Range is east of the Amur River in what country?
Russia

37. The Koni Peninsula extends into what sea bordering Sakhalin?
Sea of Okhotsk

38. The Shantar Islands belong to what country?
Russia

39. The Anzhu Islands are located north of the Lyakhovsky Islands in what country?
Russia

40. The world's deepest lake is located in Russia. Name this lake that is west of the Yablonovyy Mountain Range.
Lake Baikal

41. The West Siberian Plain is east of what major river in Russia that empties out into Ob Bay?
Ob River

42. Lake Zaysan is located in what country straddling part of the Irtysh River?
Kazakhstan

43. What country is the only doubly landlocked country in Asia?
Uzbekistan

44. The Stanovoy Mountains are located in what country bordering the Sea of Okhotsk?
Russia

45. Atyrau, formerly known as Guyev, is a city at the mouth of the Ural River in what country?
Kazakhstan

46. The Karakum Desert covers about seventy percent of what country?
Turkmenistan

47. The Aldan River is a tributary of what major Russian river?
Lena River

48. The Ishim Steppe is located in Russia and what country?
Kazakhstan

49. Pavlodar and Kostanay are cities in what country home to the Tobol River?
Kazakhstan

50. Lake Shalkar is east of the Ural River in what country whose second largest region is the Aktobe Region?
Kazakhstan

51. The Garabogazkol Aylagy forms a lagoon of what lake in Central Asia?
Caspian Sea

52. The Turan Lowland is located in Kazakhstan, Uzbekistan, and what country?
Turkmenistan

53. The Kamchatka Peninsula is east of the Sea of Okhotsk in what country?
Russia

54. Chechnya and Dagestan are republics of what country bordering the Pacific Ocean?
Russia

55. Krasnoyarsk is a city on what major Russian river?
Yenisei River

56. Ashgabat is the capital of what country bordering the Caspian Sea?
Turkmenistan

57. The Pamir Mountains can be found in what country bordering China to the east and Kyrgyzstan to the north?
Tajikistan

58. Issyk-Kul, which means "warm lake", is the largest lake in what landlocked country?
Kyrgyzstan

59. Bukhara and Qarshi are major cities in what country where part of the Ustyurt Plateau can be found?
Uzbekistan

60. Lake Tengiz and Lake Balkhash are located in what country bordering the Caspian Sea?
Kazakhstan

Middle East
Saudi Arabia, Iraq, Jordan, Yemen, Oman, Syria, United Arab Emirates, Bahrain, Kuwait, Qatar, Israel, Palestine, Lebanon

1. The Burj Khalifa, the tallest tower in the world, is located in Dubai in what country?
United Arab Emirates

2. What is the capital of Israel?
Jerusalem

3. Saudi Arabia borders what sea to the west?
Red Sea

4. Yemen borders what Gulf?

Gulf of Aden

5. The Sea of Galilee is located in the northeastern part of what country?
Israel

6. Razzaza Lake, located in central Iraq, is on the edge of what desert?
Syrian Desert

7. Sharjah is a city in what country containing the Palm Islands?
United Arab Emirates

8. Jeddah is a major city in what country home to Asir National Park?
Saudi Arabia

9. Bahrain is northeast of Saudi Arabia and northwest of what country home to the city and municipality of Al Khor?
Qatar

10. Kuwait City is a port city on what gulf with the same name as an ancient empire of Iran?
Persian Gulf

11. Salalah and Muscat are cities in what country with a Hindu minority?
Oman

12. Among the countries of Israel, Jordan, and Lebanon, which one is the largest in area?
    Jordan

13. Socotra is an island belonging to what country?
    Yemen

14. Lake Urmia is located in what country bordering the Strait of Hormuz?
    Iran

15. What mountain range is north of Tehran and south of the Caspian Sea?
    Elburz Mountains

16. Namak Lake, located approximately sixty miles east of Qom, is in what country bordering the Persian Gulf to the south?
    Iran

17. The Gulf of Masirah borders the island of Masirah to the island's north. Masirah Island belongs to what country hom to the Jabal Samhan Nature Reserve and the Arabian Oryx Sanctuary?
    Oman

18. Qeshm is an island bordering the Strait of Hormuz. Qeshm belongs to what country?
    Iran

19. R'as al-Khaimah and Sharjah are major cities in what country home to the Dubai Desert Conservation Reserve?

United Arab Emirates

20. Socotra is an island in the Indian Ocean belonging to what country on the Arabian Peninsula?
Yemen

21. Sharjah is the third most populous city in what country bordering the Persian Gulf?
United Arab Emirates

22. Halab is a major city in what country miles east of Cyprus?
Syria

23. Negev is a desert region extending more than half what country bordering the West Bank and Gaza Strip?
Israel

24. The Zagros Mountains are in the western region of what country bordering the Persian Gulf?
Iran

25. The Gulf of Aden empties out to what sea bordering Oman?
Arabian Sea

26. The Empty Quarter is a desert on what major peninsula?
Arabian Peninsula

27. Wadi Rum is a site in what country west of Syria?
Jordan

28. The Qadisha Valley is located in what country bordering Israel to the south?
Lebanon

29. Mesopotamia Marshland National Park is located in what predominantly Muslim country where Arabic and Kurdish are spoken?
Iraq

30. The Persian Gulf is connected to the Gulf of Oman by what small strait?
Strait of Hormuz

31. Islam, Hinduism, and Christianity are the largest religions of what country whose most populous city is a chief port on the Persian Gulf and whose capital is named after the country?
Kuwait

32. Asir is a region west of the Rub al Khali in what country whose provinces include Makkah and Madinah?
Saudi Arabia

33. The East Azerbaijan province is one of the 31 provinces of what country?
Iran

34. The Bab el Mandeb, bordering Djibouti, separates the Red Sea from what gulf?
Gulf of Aden

35. The geographical region of Mesopotamia extends across Iraq and what country?
Syria

36. The Kavir Desert is east of Namak Lake in what country?
Iran

37. The Tuwayq Mountains are located in what country on the Arabian Peninsula straddling the Tropic of Cancer?
Saudi Arabia

38. Timna Park is located in what country bordering the West Bank and Dead Sea?
Israel

39. The Bekaa Valley is situated between the Anti-Lebanon Mountains and what mountain range in central Lebanon?
Lebanon Mountains

40. The Jordan River, which forms part of the Israel-Jordan border, has its mouth in what sea?
Dead Sea

41. Mount Nabi Shuayb, the highest point in Yemen, is located in what mountain range?
Jabal al Hijaz

42. Hamath, Idlib, and Ar Raqqah are cities in what country bordering the Mediterranean Sea?
Syria

43. Wadi Rum is a site in what country bordering the Dead Sea?

Jordan

44. The Gulf of Awaba borders what country whose capital is Amman?
Jordan

45. Busra ash Sham is a site in what country bordering the Mediterranean Sea and Lebanon?
Syria

46. Adhari Park is located in what country bordering the Persian Gulf?
Bahrain

47. Jeddah and Najran are cities in what country bordering the Red Sea?
Saudi Arabia

48. Asir National Park is a protected area in what country?
Saudi Arabia

49. What city is the capital of Qatar, in the eastern part of the country?
Doha

50. Arabic is spoken predominantly in the Middle East, but what language is spoken primarily in Iran?
Persian (Farsi is acceptable)

51. Qurayyat is a port city in what country bordering the Arabian Sea?
Oman

52. The Arabian Oryx Sanctuary is located in what country whose capital is Muscat?
Oman

53. The Kuria Muria Islands belong to what country bordering the Gulf of Oman?
Oman

54. Masirah is an island bordering the Gulf of Masirah in what country?
Oman

55. Yemen borders what country directly north?
Saudi Arabia

56. Kamaran Island is located off the coast of what country?
Yemen

57. Kingdom Center is located in Riyadh, the capital of what country?
Saudi Arabia

58. Sakakah is a city in the northern region of what country bordering Kuwait?
Saudi Arabia

59. An exclave of Oman is separated from Iran by what strait?
Strait of Hormuz

60. Mesopotamia Marshland National Park is located miles north of the Euphrates River in what country?

Iraq

61. Iran borders what major lake to the north?
Caspian Sea

62. Lake Urmia is located west of Tabriz, a major city in what country?
Iran

63. Erbil and Mosul are major cities in what country bordering Kuwait?
Iraq

64. Socotra is the largest island in what country bordering the Gulf of Aden?
Yemen

65. The al-Shaheed Monument is a war memorial on the banks of the Tigris River to honor soldiers from what country?
Iraq

66. Madain is a site in what country bordering Saudi Arabia and Jordan?
Iraq

67. The Tropic of Cancer passes through Saudi Arabia and what country bordering the Persian Gulf and whose largest city by population is Dubai?
United Arab Emirates

68. Asir National Park is located in the southwestern region of what country?

Saudi Arabia

69. Manama is the capital of what country northwest of Qatar and east of Saudi Arabia?
Bahrain

70. Atlantis, the Palm is a major hotel resort in the Middle East, in what country whose capital is Abu Dhabi?
United Arab Emirates

71. Tehran, home to the Azadi Tower, is located on the Hableh Rood River, and is the capital of what country in the Middle East?
Iran

72. Mashhad, which is located in the Kashaf River Valley, is the second most populous city in what country bordering the Persian Gulf?
Iran

73. Naqsh-e-Jahan is a UNESCO World Heritage Site located in what city in western Iran that is located on the Zayandeh River?
Isfahan

74. Kish Island, which is a tourist attraction, is known as the Pearl of the Persian Gulf and is located in what country bordering Iraq?
Iran

75. The Kann Valley, located along the Kann River, is northwest of what major city in Iran that is the largest city in the country?
Tehran

76. Qasemabadi is a dance of the Gilak people in the Gilak Province in what country bordering the Caspian Sea and the Persian Gulf?
Iran

77. Kavir National Park, near Navak Lake and the Qom River, is located at the western edge of what major desert in Iran?
Dasht-e Kavir

78. The Imam Reza Shrine, considered one of the holiest sites by Shia Muslims, is located in the city of Mashhad in what country?
Iran

79. Bushehr, a major port city on the Persian Gulf, is located in the southern part of what Middle Eastern country whose capital is Tehran?
Iran

80. What volcano in Iran in the Elburz Mountains that is the tallest volcano in Asia is located miles northeast of the capital and major city of Tehran?
Mount Damavand

81. Zarqa, the capital of the Zarqa Governorate, is located in the Zarqa River Basin near the capital city of Amman in what country?
Jordan

82. Jabal Umm ad Dami, located in the Aqaba Governorate, is near the border with Saudi Arabia and is the highest peak in what country?

Jordan

83. Dibbeen National Park and Petra National Park are located in what country in the Middle East whose capital and largest city is Amman?
Jordan

84. Queseir Amra is a well preserved desert castle that was once a fortress and is now a UNESCO World Heritage Site in what country?
Jordan

85. The Ard As Sawwan Desert is located in the eastern part of what country?
Jordan

86. Irbid is a major city in what country bordering the Sea of Galilee and the Dead Sea to the west and Saudi Arabia to the east?
Jordan

87. Jordan borders what gulf to the southwest that opens up into the Red Sea, which separates Africa from the Arabian Peninsula in Asia?
Gulf of Aqaba

88. Dabke is one of the most popular dances in what country bordering Saudi Arabia and the Gulf of Aqaba to the south and east?
Jordan

89. Mansaf is the national and traditional dish of what country in the Middle East whose capital and most populous city is Amman?

Jordan

90. Kastrom Mefa'a, which began as a Roman military camp, is now a UNESCO World Heritage Site in what country where Islam is the majority religion?
Jordan

91. Mosul, a major city on the Tigris River, is the capital of the Nineveh Province in what country?
Iraq

92. Cheekha Dar is the lowest point in what country bordering Kuwait and whose capital is Baghdad?
Iraq

93. Nasiriyah and Fallujah are cities in Iraq on what river that forms the Shatt al Arab River after its confluence with the Tigris River?
Euphrates River

94. Hatra is a large fortified city and a UNESCO World Heritage Site in what country whose largest city is Baghdad?
Iraq

95. Lake Razzaza and Lake Tharthar are two of the largest lakes in what country bordering the Persian Gulf to the southeast?
Iraq

96. Suleymaniye is a major city in Iraq located in what mountain range in the northeastern part of the country?
Elburz Mountains

97. Basra is a major city in Iraq located on the Shatt al Arab River and is the capital of what governorate?
Basra Governorate

98. Baghdad, the capital of the Abbasid Caliphate, was formerly known as the British Mandate of Mesopotamia and is now located in what country?
Iraq

99. The Ziggurat of Ur, which is located in the ancient city of Ur near Nasiriyah, was built during the early bronze age and is now located in what country?
Iraq

100. What city in Iraq is the largest city in Iraq and the second largest city in the Middle East and is the capital of the Baghdad Province in Iraq?
Baghdad

101. Tel Aviv is a city in Israel on the Mediterranean coast within what metropolitan area in Israel?
Gush Dan

102. The Baha'i World Center is a UNESCO World Heritage Site in what major city in Israel?
Haifa

103. What lake in Israel is the largest freshwater lake in the country and is located in the Jordan Rift Valley?
Sea of Galilee

104. Masada National Park is a UNESCO World Heritage Site in what country whose capital is Jerusalem?
Israel

105. The Nahal Me'arot Nature Reserve is located on the western slopes of Mount Carmel in what country?
Israel

106. Israel's Avdat National Park is located in what major desert region?
Negev Desert

107. Megiddo National Park is located at the western entrance of valley in Israel?
Jezreel Valley

108. Nazareth, the capital of the Northern District, is a city in what country bordering the Mediterranean Sea to the west?
Israel

109. The Ashdod-Nitzanim Sand Dune Nature Reserve is located on the outskirts of the city of Ashdod in what country?
Israel

110. Tel Aviv's White City, which comprises the world's largest concentration of Bauhaus buildings, is located in what country?
Israel

Turkey and the Caucasus
Azerbaijan, Turkey, Georgia, Armenia, Abkhazia, South Ossetia, Nagorno-Karabakh

1. Lake Sevan is located in what country?
   Armenia

2. Nagorno-Karabakh is a disputed region that declared independence in 1991. It is claimed by what country home to the Absheron Peninsula?
   Azerbaijan

3. The Rioni River is found in what country bordering South Ossetia?
   Georgia

4. Mtatsminda Mountain is located in T'bilisi in what country?
   Georgia

5. Cappadocia, a region including part of the Taurus Mountains, covers much of the central part of what country?
   Turkey

6. Dilijjan National Park is north of Lake Sevan in what country?
   Armenia

7. Konya and Izmir are major cities in what country bordering the Sea of Marmara?
   Turkey

8. What city is the capital of the de facto independent but unrecognized state of Nagorno-Karabakh?

Stepanakert

9. Lake Van is located in what country bordering Armenia?
   Turkey

10. Gaziantep is a major city in what country where the Taurus Mountains are found?
    Turkey

11. Istanbul is the economic, cultural, and historical center of what country bordering the Black Sea?
    Turkey

12. What city is the second most populous in Turkey?
    Ankara

13. In 1990, South Ossetia declared independence from what country?
    Georgia

14. Lake Tuz and Lake Van are located in what country?
    Turkey

15. Azerbaijan borders what lake to the east?
    Caspian Sea

16. You can find mud volcanoes in Gobustan National Park, a site near the Caspian Sea in what country?
    Azerbaijan

17. The lira is the currency of what country whose highest point is Mount Ararat?

Turkey

18. The Etchmiadzin Cathedral is located in what country?
Armenia

19. The Absheron Peninsula is located in what country bordering the Caspian Sea?
Azerbaijan

20. The Cilician Gates and Patara Beach are located in what country?
Turkey

21. The Mingachevir Reservoir is located in the western region of what country?
Azerbaijan

22. South Ossetia borders Russia to the north and what country to the south, east, and west?
Georgia

23. What city is the largest in Abkhazia and also the region's capital?
Sukhumi

24. The Taurus Mountains are located in the southern region of what country bordering the Black Sea?
Turkey

25. Gaziantep and Adana are major cities in the southern part of what country?
Turkey

26. Svaneti is a region in what country bordering Azerbaijan to the southeast?
Georgia

27. Nakhchivan is an exclave of what country bordering Nagorno-Karabakh?
Azerbaijan

28. Ts'khinvali is the capital of what disputed region recognized independent by Russia?
South Ossetia

29. Izmir is a major city in what country bordering Georgia and Armenia?
Turkey

30. Atakule is a building in Ankara, the capital of what country bordering the Mediterranean Sea and the Black Sea?
Turkey

31. Baku, the capital of Azerbaijan, is located on what major lake?
Caspian Sea

32. Mount Bazarduzu is located in the Greater Caucasus Mountain Range and is the highest point in what country?
Azerbaijan

33. The Shahdag Mountain Resort is what country's first and largest winter resort?
Azerbaijan

34. The Talysh Mountains, located in southeastern Azerbaijan and northwestern Iran, forms the northwest section of what mountain range?
Elburz Mountains

35. Part of Hirkan National Park in Azerbaijan is located in what lowland bordering the Talysh Mountains to the south?
Lankaran Lowland

36. The Yenikend Reservoir is located in the Shamkir Rayon of what country bordering the Caspian Sea?
Azerbaijan

37. Lake Aggol is a large salty lake in the Kur-Araz Lowland of what country in the Caucasus?
Azerbaijan

38. Azerbaijan is a part of the Turkic Council, which consists of how many independent states?
Six

39. Ilham Aliyev is the president of what Caucasian country bordering the Caspian Sea?
Azerbaijan

40. The Baku-Tbilisi-Ceyhan Pipeline is a long crude oil pipeline from the Caspian Sea to what other sea?
Mediterranean Sea

41. The Magomayev Azerbaijan State Philharmonic Hall is the main concert hall, built in 1912, and is located in what Azerbaijani city?

Baku

42. The Ismailiyya Palace is located on Istiglaliyyat Street in what city in Azerbaijan?
Baku

43. The European Grand Prix is a Formula One event, and will be held in 2016 in the city of Baku in what country?
Azerbaijan

44. What city in Azerbaijan is known by the nickname as the "City of Winds"?
Baku

45. The Palace of the Shirvanshahs is a UNESCO World Heritage Site and is located in what city in Azerbaijan?
Baku

46. What city in Azerbaijan was formerly named as Elizabethpol while part of the Russian Empire?
Ganja

47. The 2015 European Games were held in what city on the Absheron Peninsula and Baku Bay in Azerbaijan?
Baku

48. Lake Boyukshor is the largest lake by area on what peninsula in Azerbaijan?
Absheron Peninsula

49. The Azeri TV Tower is a concrete telecommunications building that is the tallest structure in what country?
Azerbaijan

50. The Ghazanchetsots Cathedral is located in the city of Shusha in the disputed region of Nagorno-Karabakh, which is part of what Caucasian country?
Azerbaijan

51. The Fountains Square is a public square in the downtown area of what major Azerbaijani city?
Baku

52. The Bibi-Heybat Mosque is a historical and cultural structure in what city in Azerbaijan?
Baku

53. The Heydar Aliyev Center is a building complex designed by the architect Zaha Hahid in Baku in what country?
Azerbaijan

54. The Maiden Tower, also known locally as Giz Galasi, is a UNESCO World Heritage Site in what city in Azerbaijan?
Baku

55. Novruz is an Iranian festival celebrating the New Year and is celebrated in what country bordering Russia, Georgia, and the Caspian Sea?
Azerbaijan

56. Mount Aragats is the highest peak in what landlocked Caucasian country?
Armenia

57. What lake, the largest lake by area in Armenia, is located in the Sevan Basin in the eastern part of Armenia?
Lake Sevan

58. The Ararat Plain lies at the foot of what mountain in Armenia?
Mount Aragats

59. The Aras River cuts what plain in half, the northern part in Armenia and the southern part in Iran and Turkey?
Ararat Plain

60. Gyumri is the second most populous city in what mountainous landlocked Caucasian country?
Armenia

61. The Akhuryan River is a 130-mile left bank tributary of what major river in Armenia and the Caucasus geographical region?
Aras River

62. Armenian is a language and also a distinct branch of what major worldwide language family?
Indo-European Languages

63. The Armenian dram, established in 1993, replaced what other currency that had been used in Armenia?
Ruble

64. Vanadzor is a machine-building city and a center of chemical industry in what landlocked Caucasian country?
Armenia

65. Gyumri and Vanadzor are major industrial cities in what country?
Armenia

66. What country is Armenia's major import source and export destination?
Russia

67. Armenia is divided into how many provinces, also known as oblasti?
Ten

68. The Akhurian River in Armenia originates in what lake?
Lake Arpi

69. Lake Arpi, a Ramsar Convention Protected Site, is located in what province in Armenia?
Shirak Province

70. The Spandaryan Reservoir was built around 1980 in the Vorotan River Basin in Armenia and is connected to what river?
Arpa River

71. The Tatev Monastery is located in what southeastern province of Armenia?
Syunik Province

72. Amberd is a 7th century fortress on the slopes of Mount Aragats, at the confluence of the Arkashen River and what other river in Armenia?
Amberd River

73. The Etchmiadzin Cathedral, the mother church of the Armenian Apostolic Church, is located in what city in Armenia?
Vagharshapat

74. The Azat River, located in the Kotayk Province of Armenia, is on the western slope of what mountain range?
Geghama Mountains

75. The Zangezur Mountains, historically known as the Syuniq Mountains, forms the border between the Azerbaijani exclave of Nakhchivan and what Armenian province?
Syunik Province

76. Arevik National Park is located in the southern Syunik Province of what country in the Caucasus?
Armenia

77. Sevanavank is a monastic complex located on the shores of what lake in Armenia in the Gegharkunik Province?
Lake Sevan

78. Victoria Park is a city park located in the Kanaker-Zeytun District of Yerevan, the capital of what Caucasian country?
Armenia

79. The Zvartnots Cathedral is located in the city of Armavir in what country?
Armenia

80. The Cafesjian Museum of Arts is located in Yerevan, the capital of what mountainous landlocked Caucasian country?
Armenia

81. The Katoghike Church is located in the Kenton District of what major city in Armenia that is also the capital of Armenia?
Yerevan

82. Armavir and Gyumri are major cities in what mountainous landlocked Caucasian country whose capital and largest city is Yerevan?
Armenia

83. What is the largest river in Transcaucasia, a region that consists of the countries of Armenia, Georgia, and Azerbaijan?
Kura River

84. The Debed River contains the lowest point in what country in the Caucasus region?
Armenia

85. The Aras River forms the entire border between Armenia and what country to the east?
Iran

86. The Aragatsotn Province and Tavush Province are located in what country whose largest lake by area is Lake Sevan?
Armenia

87. Yerevan, the capital of Armenia, is located on the banks of what major river in Armenia?
Hrazdan River

88. The Saint Gregory the Illuminator Cathedral is located in the Kenton District of what city in Armenia?

Yerevan

89. Tamanyan Street, which links the Yerevan Cascade to Moscow Street, is home to the Cafesjian Museum of Art in what country?
Armenia

90. The Yerevan Opera Theater, which was designed by the architect Alexander Tamanian, is located in what country?
Armenia

91. The Yerevan Cascade is a huge stairway in the city of Yerevan in what country?
Armenia

92. The Republic Square is a large central town square in the city of Yerevan in what country in the Caucasus region?
Armenia

93. Goshavank is a monastery located in the village of Gosh in what province in Armenia?
Tavush Province

94. The Sanahin Monastery is located in the Lori Province of what landlocked country in the Caucasus?
Armenia

95. The Saint Hripsime Church, one of the oldest surviving churches in Armenia, is located in what city?
Vagharshapat

96. The Dalma Garden Mall is located in the capital city of Yerevan in what country?

Armenia

97. Borjomi-Kharagauli National Park is a protected area southwest of the city of Tbilisi in what country?
Georgia

98. Mtirala National Park is a protected area in the Adjara region of what country whose capital is Tbilisi?
Georgia

99. Kutaisi, the legislative capital of Georgia, is the capital of what region in Georgia?
Imereti

100. Batumi, an important port city and commercial center, is a seaside city on the Black Sea that is the capital of what region in Georgia?
Adjara

101. Kutaisi, the capital of the Imereti region in Georgia, is located on what river?
Rioni River

102. What major city in Georgia was formerly known as Tiflis and is located on the banks of the Kura River?
Tbilisi

103. Mount Shkhara is located in the Svaneti region of what country whose largest cities are Tbilisi, Kutaisi, and Batumi?
Georgia

104. The Kodori Valley is located in what disputed region in Georgia bordering Russia to the north?

Abkhazia

105. The Kabargin Oth Group is a group of volcanoes in what disputed region of Georgia whose capital is Tskhinvali?
South Ossetia

106. Paravani Lake, Kartsakhi Lake, and Paliastomi Lake are the three largest lakes by area in what country?
Georgia

A Competitor's Compendium to the Geography Bee

Europe

Northern Europe
Sweden, Finland, Denmark, Norway, Iceland, Estonia, Latvia, Lithuania, Aland Islands, Faroe Islands, Svalbard, Jan Mayen

1. Maanselka is a plateau region in Russia, Norway, and what Nordic country?
   Finland

2. St. Petersburg is a port city on what gulf bordering the Baltic Sea?
   Gulf of Finland

3. The Store Baelt is a strait in what country whose lowest point is Lammefjord?
   Denmark

4. Lake Pskov is on Russia's border with what country?
   Estonia

5. Juozapines is the highest point in what country containing the Nemunas River?
   Lithuania

6. Lake Vanern and Lake Vattern are lakes in what country with the regions of Norrland and Svealand?
Sweden

7. Tivoli is a world-famous amusement park in downtown Copenhagen in what country?
Denmark

8. Tallinn is a city that is home to about a third of the population of what country bordering the Gulf of Finland?
Estonia

9. The Faroe Islands, where the capital is Torshavn, is an island territory of what country?
Denmark

10. Nokia Corporation's world headquarters are located in Espoo, the second largest city in Uusimaa, a region in what country with a literacy rate of 100%?
Finland

11. Vatnajökull National Park, established in 2008, is located in what country?
Iceland

12. Lake Vattern is a major lake in what country containing the cities of Norrkoping and Uppsala?
Sweden

13. Vaasa and Tampere are major cities in what country bordering the Gulf of Bothnia?

A Competitor's Compendium to the Geography Bee

Finland

14. Vardo is a city in what country bordering the North Sea and Norwegian Sea?
Norway

15. Tampere-Pirkkala Airport is located in what country containing the Oulujoki River?
Finland

16. Rosenborg Castle is a Renaissance building located in what city in Denmark on the eastern coast of the island of Zealand?
Copenhagen

17. Pokaini Forest is located in what country bordering the Irbe Strait to the north?
Latvia

18. The Seaside Open-Air Museum is located in Ventspils in what country?
Latvia

19. Skagerrak and Kattegat are straits bordering what country whose capital is Copenhagen?
Denmark

20. Jutland is located in what country whose main language is Danish?
Denmark

21. The Western Dvina River is located in Belarus, Russia, and what country?
Lithuania

22. Arhus is a city on the eastern coast of what Danish peninsula?
Jutland

23. The Fortress of Suomenlinna is located in what country bordering the Gulf of Bothnia?
Finland

24. Lake Malaren is located in what country whose major cities include Malmo, Helsingborg, and Goteborg?
Sweden

25. What island is the largest in Sweden, bordering the Baltic Sea?
Gotland

26. The Norse Folk Museum is located in Oslo, the capital of what country?
Norway

27. Blue Lagoon is located in what country whose highest point is Hvannadalshnukur at 6,923 feet?
Iceland

28. Daugavpils and Riga are major cities in what country bordering the Irbe Strait?
Latvia

29. Vilnius is the capital of what country bordering Courland Lagoon?
Lithuania

30. Oulanka National Park is located in the northern region of what country where the cities of Oulu and Tampere can be found?
Finland

31. Espoo and Turku are major cities in what country bordering Russia?
Finland

32. Jostedalsbreen Glacier is located in what country bordering Skagerrak?
Norway

33. Foteviken Viking Reserve is located in what country where Lake Vanern can be found?
Sweden

34. Kebnekaise and Sarjektjakko are peaks in the northern region of what country?
Sweden

35. The Faroe Islands are occupied by what country bordering Germany?
Denmark

36. What glacier in Iceland is the largest in Europe?
Vatnajokull

37. A Legoland theme park is located in the city of Billund in what country?
Denmark

38. Bornholm is an island belonging to what country bordering the North Sea?
Denmark

39. Hiiumaa is the second largest island after Saaremaa in what country whose capital is Tallinn?
Estonia

40. The Gulf of Riga borders Latvia and what other country?
Estonia

41. Kaunas is a major city in what country bordering Latvia to the north?
Lithuania

42. The Stockholm Archipelago is located east of what country bordering the North Sea and the Baltic Sea?
Sweden

43. Geirangerfjord is a protected area in what country where the cities of Trondheim and Stavanger can be found?
Norway

44. Ovre Anarjohka National Park is located in Finnmark County in what country?
Norway

A Competitor's Compendium to the Geography Bee

45. Lemmenjoki National Park, one of the largest in Europe, is the largest by area in what country?
Finland

46. Falster, Lolland, Fyn, Zealand, and Langeland are islands in what country home to the city of Aarhus?
Denmark

47. Malmo is a major city in what country that borders two other countries, Finland and Norway?
Sweden

48. The Oresund Bridge is located in what country besides Denmark bordering the Kattegat?
Sweden

49. Naantali is a city a few miles west of Turku in what country home to the Aland Islands?
Finland

50. The Salpaus Ridge is located in what country containing the cities of Oulu, Tampere, Espoo, Turku, and Helsinki?
Finland

Western Europe
United Kingdom, France, Germany, Spain, Italy, Portugal, Ireland, Netherlands, Austria,

A Competitor's Compendium to the Geography Bee

Switzerland, Belgium, Luxembourg, Andorra, Malta, Liechtenstein, San Marino, Monaco, Vatican City, Isle of Man, Gibraltar, Jersey, Guernsey

1. Clew Bay and Dingle Bay both form part of the western and southwestern coasts of what country?
Ireland

2. The Black Forest is a mountainous region in the southwestern part of what country?
Germany

3. Plateau de Langres is located in what country where the city of Avignon is situated on the confluence of the Rhone and Durance Rivers?
France

4. The Moselle River forms part of Luxembourg's border with what country?
Germany

5. Ponta do Pico, in the Azores Islands, is the highest point in what country?
Portugal

6. Gozo is the second largest island in what country bordering the Mediterranean Sea?
Malta

7. Rugen is an island off the northeastern coast of what country bordering the Baltic Sea?
   Germany

8. In 2003, what European city became home to the United Nations' permanent court for prosecuting war criminals?
   The Hague

9. Brenner Pass connects Austria to what country bordering the Gulf of Gaeta and the Strait of Messina?
   Italy

10. Pomerania is a region extending into Poland in what country?
    Germany

11. Helgolander Bay feeds into what sea?
    North Sea

12. 11 communes make up the Principality of Liechtenstein. This doubly-landlocked country borders Austria and what other nation?
    Switzerland

13. Amsterdam is the chief port city on Ijsselmeer Lake in what country whose provinces include Friesland and Overijissel?
    Netherlands

14. The mouth of the Guadiana River is situated in what gulf off the coast of Portugal?
    Gulf of Cadiz

15. Mecklenburger is a bay forming part of the northern coast of what country?
Germany

16. The Orkney Islands belong to what country?
United Kingdom

17. The Po River is the longest river in what country?
Italy

18. The Cambrian Mountains are in the southern region of what country in the British Isles?
United Kingdom

19. The Celtic Sea borders what country to the north whose capital is Dublin?
Ireland

20. The Tiber River is located in what country bordering the Gulf of Taranto?
Italy

21. The Loire River is located in what country bordering the Bay of Biscay?
France

22. The Balearic Islands belong to what country bordering Andorra?
Spain

23. Munich is a city in what country bordering Switzerland and Austria?
Germany

24. Corsica and Sardinia are islands in the Mediterranean Sea. Corsica belongs to France and Sardinia belongs to what country?
Italy

25. Monaco borders what country containing the cities of Limoges and Strasbourg?
France

26. Dundalk Bay feeds into what sea bordering the United Kingdom and Ireland?
Irish Sea

27. The Adige River, whose source is in the Alps, has its mouth in what gulf with the same name as a famous Italian city?
Gulf of Venice

28. Mont Blanc de Courmayeur is on Italy's border with what country?
France

29. Dufourspitze is a peak on Switzerland's border with what country bordering the Ligurian Sea and the Gulf of Trieste?
Italy

30. Cardigan Bay forms part of the western coast of what country separated from Ireland by the St. George's Channel?
United Kingdom

31. The Aran Islands are in what country bordering the Celtic Sea?
Ireland

32. Majorca, Minorca, and Iviza are islands in the Balearic Sea belonging to what country?
Spain

33. The Gulf of Valencia is in the Balearic Sea east of what country?
Spain

34. The Rock of Cashel is located in Tipperary in what country bordering Dunmanus Bay and Bantry Bay?
Ireland

35. Rouen and Limoges are cities in what country containing the mouth of the Gironde River?
France

36. Funchal is the capital of what island group belonging to Portugal?
Madeira Islands

37. Las Palmas is the capital of what island group belonging to Spain?
Canary Islands

38. Pomerania is a region in Germany and Poland south of what sea?
Baltic Sea

39. Schleswig-Holstein, Thuringia, and Lower Saxony are states in what country containing the Mosel and Main Rivers?
Germany

40. Catalan is the main language spoken in what country whose capital city has approximately 23,000 people?
Andorra

41. What country bordering the Bay of Biscay is the largest in Western Europe?
France

42. Volkswagen and Daimler are well known industrial names from what country bordering the Netherlands?
Germany

43. German is the official language of what country whose capital is Vienna?
Austria

44. Brussels is the headquarters for the European Union and the North Atlantic Treaty Organization. This city is the capital of what country whose currency is the euro?
Belgium

45. Potsdam is the capital of what German state in the northeast bordering Poland?
Brandenburg

46. Schönbrunn Palace was the home of the Habsburg emperors in the 18th and 19th centuries. This palace is in what landlocked country bordering Lake Constance?
Austria

47. The Tagus River flows through what capital city on the Iberian Peninsula?
Lisbon

48. Lyme Bay borders what country to the north?
United Kingdom

49. Guernsey and Jersey are territories west of France. These territories belong to what country?
United Kingdom

50. The Keukenhof Flower Garden is located in what city in the Netherlands?
Lisse

51. The Bock Casemates is a fortress in what country?
Luxembourg

52. The Saone River is a tributary of what river whose mouth is in the Gulf of Lion?
Rhone River

53. Limousin, a region in France, is west of what mountain range located in southern France?
Massif Central

54. The stone carvings of Val Camonica in the province of Brescia is home to one of the largest collections of prehistoric petroglyphs in the world. This collection of stone carvings has been made into a UNESCO World Heritage Site and is in what country?
Italy

55. Monte San Giorgio is a site overlooking Lake Lugano in two countries. Name these countries.
Italy and Switzerland

56. Italian, German, French, and Slovene are four of the major languages spoken in what country bordering the Ligurian Sea?
Italy

57. The Gironde Estuary feeds into what Bay bordering France?
Bay of Biscay

58. Mont Agel is the highest point in what French-speaking country?
Monaco

59. Foreste Casentinesi National Park is located in what country bordering France to the west?
Italy

60. Punta La Marmora is the highest peak in the Gennargentu Mountain Range in what country?
Italy

61. Albert II is the prince of what country bordering France?
Monaco

62. Latin belongs to what branch or subfamily of the Indo-European language family?
Italic Languages

63. The Gulf of Valencia feeds into what sea?
Balearic Sea

64. The Ebro River has its source in what mountain range in northern Spain?
Cordillera Cantabrica

65. Cagliari is the capital of what Italian island that is the second largest in the Mediterranean Sea?
Sardinia

66. The Bernese Alps and the Rhaetian Alps are located in what country bordering Italy to the south?
Switzerland

67. Belem Tower is a fortress built in the early 16th century in what country?
Portugal

68. The Alhambra is a palace and fortress in what city in Spain?

Granada

69. Ducie Island, in Polynesia, belongs to what European country?
United Kingdom

70. The Bavarian Alps is located along Germany's border with what country?
Austria

71. Connemara is a lowland region north of Galway Bay in what country?
Ireland

72. What country is the world's leading producer of cork?
Portugal

73. Ceuta and Melilla, although in Morocco, belongs to what country to the north?
Spain

74. Schaan is a municipality of what country bordering Switzerland and Austria?
Liechtenstein

75. Lake District National Park is located in what country bordering the North Sea?
United Kingdom

76. Geneva is located at the confluence of the Arve and what other river?
Rhone River

77. Stoclet Palace is a UNESCO World Heritage Site in Brussels in what country bordering France?
Belgium

78. The Cotentin Peninsula is bordered by what gulf?
Gulf of Saint-Malo

79. Tarragona, Zaragoza, and Pamplona are cities in what country bordering the Alboran Sea?
Spain

80. Antwerp is a major city in what country that is famous for its chocolates?
Belgium

Southern Europe
Greece, Romania, Bulgaria, Serbia, Croatia, Albania, Bosnia and Herzegovina, Macedonia, Slovenia, Montenegro, Kosovo, Cyprus, Northern Cyprus, Akrotiri and Dhekelia

1. Bobotov Kuk is the highest point in what country bordering the Ionian Sea?
Montenegro

2. Lake Ohrid is situated on the border of what two countries?
Albania and Macedonia

3. Moldoveanu Peak is the highest point in what country home to part of the Great Hungarian Plain in its western region?
Romania

4. Corfu is an island off the northwestern coast of what country?
Greece

5. The Danubian Plain, located east of the Timok River and west of the Black Sea, is located in what country home to part of the Strandzha Mountains?
Bulgaria

6. Cephalonia is an island off the western coast of what country?
Greece

7. The Gulf of Trieste is west of what country whose capital, Ljubljana is a chief port city on the Sava River?
Slovenia

8. Lake Prespa is on the edge of the Pindus Mountains on what country's border with Albania and Macedonia?
Greece

9. What country whose capital is Sarajevo exports metals, clothing, and wood products?
Bosnia and Herzegovina

10. Croatia declared its independence from what former country in 1991?
    Yugoslavia

11. Bohemia and Moravia are regions in what country home to Central Europe's oldest university, Charles University, which was founded in Prague in 1348?
    Czech Republic

12. A euro crisis devastated the economy of what country where democracy was born in the fifth century B.C.?
    Greece

13. The Belogradchik rocks are located in what country bordering Turkey?
    Bulgaria

14. Rila National Park is located in what country bordering Romania to the north?
    Bulgaria

15. Podgorica is the capital of and the largest city in what country containing part of the Dinaric Alps?
    Montenegro

16. Belgrade is located on the confluence of the Danube and what other river?
    Sava River

17. Walachia is a region south of what mountain range in Romania?
    Transylvanian Alps

18. The Vardar River is the lowest point in what country bordering Greece and Bulgaria?
Macedonia

19. Greek, Vlach, and Romani are minority languages in what country whose government is a parliamentary democracy and whose currency is the lek?
Albania

20. Sarajevo is the capital of what county bordering Montenegro and Croatia?
Bosnia and Herzegovina

21. The Karavasta Lagoon is the largest lagoon in the Adriatic Sea and also in what country bordering Macedonia?
Albania

22. Stobi was an ancient town in Paeonia in what country bordering Kosovo to the north?
Macedonia

23. The Neretva River is located in what country whose capital is Sarajevo?
Bosnia and Herzegovina

24. Samaria National Park is located on what island in Greece bordering the Sea of Crete?
Crete

25. The island of Samos is west of Turkey and belongs to what country?

Greece

26. Serbia borders what country to the southwest that shares Lake Scutari with Albania?
    Montenegro

27. Lake Prespa is shared by Macedonia, Albania, and what country?
    Greece

28. The Belogradchik Rocks are located in what country bordering Romania?
    Bulgaria

29. The island of Corfu is southwest of Albania and belongs to what country?
    Greece

30. Transylvania is a geographic region in what country whose largest city and capital is Bucharest?
    Romania

31. The Parthenon is located in what city in mainland Greece that is the most populous in the country?
    Athens

32. Santorini is an island north of the Sea of Crete in what country?
    Greece

33. Morphou Bay borders what country to the south that is home to the Karpass Peninsula, one of its most recognizable geographical features?
Cyprus

34. The Karpas Peninsula is located in what disputed region?
Northern Cyprus

35. Mount Olympus is located in what country bordering the Thermaic Gulf?
Greece

36. The Cyclades are located in what country?
Greece

37. Cape Apostolos Andreas is located on the Karpass Peninsula on what Mediterranean island?
Cyprus

38. Mount Attavyros is a peak on what Greek island situated south of Turkey?
Rhodes

39. The Troodos Mountains are located in what country?
Cyprus

40. Cape Greco is located in what country bordering Episkopi Bay and Akrotiri Bay?
Cyprus

41. Split is a major city in what country whose capital is Zagreb?

Croatia

42. Kosovo borders Macedonia, Albania, Serbia, and what country?
Montenegro

43. What is the capital of Montenegro, located in the southern region of the country a few miles north of Lake Scutari?
Podgorica

44. Lake Ohrid is located in two countries. Name these countries north of Greece.
Albania and Macedonia

45. The Apuseni Mountains are located in the region of Transylvania what country bordering Moldova?
Romania

46. Bulgaria borders what sea to the east?
Black Sea

47. Cape Kormakitis is located in what disputed region bordering Famagusta Bay?
Northern Cyprus

48. Skyros and Lemnos are islands in what country bordering the Ionian Sea?
Greece

49. Crete, Rhodes, Lemnos, Zante, and Cephalonia are islands belonging to what country bordering the Aegean Sea and the Mediterranean Sea?

Greece

50. Mount Olympus is the highest peak in what mountain range in Cyprus that is the largest mountain range in Cyprus?
Troodos Mountains

51. The Pedieos River, the longest river in Cyprus, empties out into Famagusta Bay and originates in what major mountain range, near the Machairas Monastery?
Troodos Mountains

52. The Machiras Monastery, located near the village of Lazanias, is located in what country whose capital is Nicosia?
Cyprus

53. The Kouris Dam is located in the Limassol District in what country bordering the Mediterranean Sea on all sides?
Cyprus

54. Larnaca International Airport is located in the city of Larnaca in what Mediterranean country whose capital is Nicosia?
Cyprus

55. Limassol and Stravolos are major cities in what country whose capital and largest city is Nicosia?
Cyprus

56. The Walls of Nicosia are a series of defensive walls surrounding the city of Nicosia, the capital of what country?
Cyprus

57. The Kyrenia Range is located in what country where Nicosia and Kyrenia are major cities?
Cyprus

58. Famagusta Bay borders what country that is home to the Machiras Monastery and Larnaca International Airport?
Cyprus

Central Europe
Poland, Czech Republic, Slovakia, Hungary

1. Gerlachovsky Stit is located in the Carpathian Mountains and is the highest point in what country home to the Vah and Hornad Rivers?
Slovakia

2. Budapest is the capital of what landlocked country?
Hungary

3. The Krkonoše Mountains, whose highest point is Snezka, are located on what country's border with Poland?
Czech Republic

4. The Kremnica Mint is a factory in what country whose peaks include Gerlach and Rysy?
Slovakia

5. Slovak Karst National Park is in what country?
Slovakia

6. Lake Balaton, also known as the Hungarian Sea, is the largest lake in what country?
Hungary

7. Castle Lucen is a landmark in the northern region of what landlocked country?
Czech Republic

8. Sand dunes are common in Slowinski National Park in what country?
Poland

9. Budapest is a city in Hungary on what major European river that forms part of the border between Hungary and Slovakia?
Danube River

10. The Velvet Revolution, in 1989, was an uprising against the government of what former nation that has now split into multiple countries?
Czechoslovakia

11. The highest point in the Carpathian Mountains is located in what country?
Slovakia

12. The Northern European Plain Stretches from Germany all the way to what country?
 Russia

13. Slowinski National Park is located in the northern region of what country bordering the Baltic Sea?
 Poland

14. Pieniny National Park is located in what country?
 Poland

15. Moravia is a region in what country bordering Slovakia to the east?
 Czech Republic

16. Nyiregyhaza is a major city in what country?
 Hungary

17. Budapest is located on what major European river?
 Danube River

18. Slovak Karst National Park is a site in what country?
 Slovakia

19. The Dunajec River flows through Pieniny National Park in what country?
 Poland

20. You can visit the ruins of Spis Castle, one of the largest castles in Europe. This castle is located in what country?
 Slovakia

21. Neusiedler Lake borders what country to the south?
Hungary

22. What country borders the Kaliningrad Oblast to the north?
Poland

23. Castle Lucen is located in what country whose capital is Prague?
Czech Republic

24. Malbork Casle is a site in Malbork in what country?
Poland

25. Gerlach is a peak in the northern region of what country?
Slovakia

26. The Kremnica Mint is located in the city of Kremnica in what country?
Slovakia

27. The Wieliczka Salt Mine is located in what country?
Poland

28. The Prague Astronomical Clock is located in what country?
Czech Republic

29. The Vah and Hron Rivers are located in what country whose capital is Bratislava?
Slovakia

30. The cities of Miskolc and Szekesfehervar are located in what country whose largest lake is Lake Balaton?
Hungary

Eastern Europe and European Russia
Ukraine, Belarus, Moldova, Transnistria, European Russia

1. The Chersky Mountain Range is located in the northeastern region of what country home to part of the Caspian Depression?
Russia

2. The Rybinsk Reservoir is located in western Russia and is fed by what river?
Volga River

3. Crimea is a peninsula in what country bordering Belarus and Moldova?
Ukraine

4. Svernaya Zemlya is an island group off the coast of the Taymyr Peninsula bordered by the Kara Sea and what other sea?
Laptev Sea

5. The Gorki Reservoir is on what river that is the longest in Europe?

Volga

6. Kharkiv and Donetsk are major cities in what country containing the Bug and Donets Rivers?
Ukraine

7. The Oka-Don Plain is located east of what upland in Russia?
Central Russian Upland

8. The Pethara Basin is located west of what mountain range?
Ural Mountains

9. The Dnieper and Black Sea Lowlands are located in what country?
Ukraine

10. The Onega River empties out into what sea that is an inlet of the Arctic Ocean?
White Sea

11. The Pripyat Marshes are located in what country whose capital is Minsk?
Belarus

12. The Kola Peninsula is located in what country bordering Finland?
Russia

13. What is the largest lake entirely in Europe and in what country is it located in?
Lake Ladoga, in Russia

14. The ruble is the official currency of what landlocked country that was formerly part of the Soviet Union bordering the Ukraine?
Belarus

15. Pirin National Park is a UNESCO World Heritage Site in what country that is one of the poorest in the European Union?
Belarus

16. Braslau Lakes National Park is in the northwest region of what country bordering Lithuania?
Belarus

17. The Gulf of Taganrog is southeast of what country containing the Kremenchuk Reservoir?
Ukraine

18. Transnistria, since the breakup of the Soviet Union, has been struggling for independence from what country?
Moldova

19. Mir Castle is an attraction in what country bordering Latvia?
Belarus

20. Karkinit Gulf is west of what major peninsula in the Ukraine?
Crimean Peninsula

21. The Dnieper River's mouth is located in what sea south of Eastern Europe?
Black Sea

22. The Pinsk Marshes are located west of the Dnieper River in the southern region of what country containing the Western Dvina River?
Belarus

23. The city of Murmansk is located on what peninsula in Russia?
Kola Peninsula

24. The Motherland Statue was built in 1967 to remember the Battle of Stalingrad. This statue is located in Volgograd in what country?
Russia

25. The Volga River Delta feeds into what sea bordering Russia?
Caspian Sea

26. In October 2015, Alexander Lukashenko was re-elected for his fifth term in office in what country?
Belarus

27. The Kakhovka Reservoir is located in what country bordering the Karkinit Gulf?
Ukraine

28. The Volyn-Podolian Upland is located in what country bordering Slovakia and Poland?
Ukraine

29. Lake Onega is a major lake in what country bordering Latvia?
    Russia

30. Ukrainian and Russian minorities are struggling for independence from Moldova in what disputed region?
    Transnistria

31. Kharkiv and Donets'k are major cities in the eastern region of what country where the Donets Basin can be found?
    Ukraine

32. Kandalaksha Bay borders the White Sea to the east and what major peninsula to the north?
    Kola Peninsula

33. Kirov and Syktyvkar are cities in what country bordering the Sea of Azov?
    Russia

34. The Tsimlyansk Reservoir is located in what country bordering Ukraine?
    Russia

35. Braslau Lakes National Park is located in the northern part of what country?
    Belarus

36. Kremenchuk is a city in what country where the Dnieper River can be found?
    Ukraine

37. Bender and Comrat are cities in what country bordering Ukraine?
Moldova

38. Tiraspol is the capital of what disputed region trying to reach independence?
Transnistria

39. Barysaw and Mahilyow are major cities in what country bordering Poland and Russia?
Belarus

40. Lake Lacha, Lake Vozhe, and the Kanin Penisnula are located in what country bordering Chesha Bay to the north?
Russia

Africa

Northeast Africa
Egypt, Sudan, Ethiopia, Somalia, Eritrea, Djibouti, South Sudan, Kenya, Uganda, Burundi, Rwanda, Tanzania, Somaliland

1. Mount Kenya is the second highest mountain in Africa after what mountain in Tanzania?
   Mount Kilimanjaro

2. Lake Ch'ew Bahir borders Kenya and what country?
   Ethiopia

3. The Selous Game Reserve, one of the largest faunal reserves in the world, protects lions, elephants, and other wild animals. This reserve is located in what country containing Lake Manyara and Lake Eyasi?
   Tanzania

4. The Ilemi Triangle is administered by Kenya but claimed by what nation?
   South Sudan

5. Meroe, the ancient capital of the Kingdom of Kush, is located in what country home to the Sanganeb Lighthouse and bordering the Red Sea?
Sudan

6. The Blue Nile and White Nile rivers met at the major city of Khartoum in what country?
Sudan

7. Lake Burundi borders what major African lake to the southwest?
Lake Tanganyika

8. Day Forest National Park is located in what country bordering Lake Abbe?
Djibouti

9. Baardheere is a city situated about two hundred miles west from Mogadishu in what country?
Somalia

10. Sinbusi Beach, a Somalian tourist attraction, is near what major city?
Marka

11. THe Laas Geel Caves contains some of Africa's best-preserved cave paintings, showing humans and animals. These caves are located in what country bordering Kenya?
Somalia

12. Volcanoes National Park is located in what country bordering Lake Kivu?
    Rwanda

13. Entebbe is a city in Uganda on a peninsula jutting out into what lake?
    Lake Victoria

14. The Ssese Islands belong to what country home to Lake George and the Kazinga Channel?
    Uganda

15. You can find leopards, cheetahs, and lions on the grasslands of Kidepo Valley National Park is the northeastern region of what country that shares Lake Albert with the Democratic Republic of the Congo?
    Uganda

16. The Awash River, in Ethiopia, is in the Great Rift Valley. This river, which has its source in the Ethiopian Highlands, has its mouth in what lake?
    Lake Abbe

17. Mount Catherine is located in what country with the city of Shubra al Khaymah?
    Egypt

18. The Bab el Mandeb, which connects the Gulf of Aden to the Red Sea, separates Yemen from what country?
    Djibouti

19. Dinder National Park, bordering the Rahad River to the north, is also a biosphere reserve, and is located in the southeastern region of what country?
Sudan

20. Fasil Ghebbi is a fortress-city in what country where Lake Zway and Lake Abaya can be found?
Ethiopia

21. Mount Kenya is closest to what line of latitude that passes through the cities of Meru and Nyahururu?
Equator

22. Lake Amboseli is on Kenya's border with what country?
Tanzania

23. The Serengeti Plain is located in the northern part of what country?
Tanzania

24. Lake Kyoga is located in what country bordering Rwanda?
Uganda

25. The Nubian Desert is east of the second and third cataracts of what river?
Nile River

26. Cape Guardafui, also known as Ras Asir, is located in the Bari Region of Puntland in what country?
Somalia

27. Maasai Mara National Reserve is located in what country bordering South Sudan?
Kenya

28. Beledweyne is a major city in what country bordering Ethiopia and Kenya?
Somalia

29. Yangudi Rassa National Park and Gambela National Parks are located in what landlocked country bordering Somaliland?
Ethiopia

30. Ungama Bay, in the Indian Ocean, borders what country containing the Gede Ruins?
Kenya

31. Lake Manyara National Park is in what country that is home to Lake Natron?
Tanzania

32. Simien Mountains National Park is located in what country whose capital is Addis Ababa?
Ethiopia

33. Valley of the Kings is a site in what country bordering Foul Bay?
Egypt

34. The Achwa and Katonga Rivers flow through what country?
Uganda

35. The Massawa War Memorial is located in what country bordering the Red Sea?
Eritrea

36. The Qattara Depression is located in what country containing the Western Desert and Libyan Desert?
Egypt

37. The Aswan Botanic Gardens are located in Aswan in what country?
Egypt

38. As part of a coming-of-age ritual, Maasai boys in Kenya must kill what wildcat that can be found in Kora National Park?
Lion

39. The Bahr el 'Arab River is located in what country that was formally established in 2011?
South Sudan

40. The Hanish Islands in the Red Sea are claimed by Yemen and what country?
Eritrea

41. The Danakil Depression is the lowest point in what country?
Ethiopia

42. Shimbiris is a peak in what country bordering the Gulf of Aden?
Somalia

43. Lake Nasser is located in what country whose capital is Cairo?
Egypt

44. Ras Dashen is a peak in what country containing the Ethiopian Highlands?
Ethiopia

45. Pemba Island belongs to what country bordering Lake Tanganyika?
Tanzania

46. The Chalbi Desert, situated east of Lake Turkana, is located in what country bordering Uganda?
Kenya

47. The Serengeti Plain in Kenya extends into what country bordering Rwanda and Burundi?
Tanzania

48. The Toshka Lakes are in the southern region and in the New Valley Governorate of what country bordering the Red Sea?
Egypt

49. Pemba Island, known as the "Green Island" in Arabic, is situated north of what island in Tanzania?
Zanzibar

50. The Dahlak Archipelago, near Massawa, belongs to what country bordering Djibouti?
Eritrea

51. The Iwembere Steppe and Maasai Steppe are located in what country?
Tanzania

52. Lake Rukwa, an endorheic lake in the Rukwa Valley that lies midway between Lake Malawi and Lake Tanganyika, is located in what major African valley?
Great Rift Valley

53. The Eritrean-Ethiopian war, whose basis was the dispute over the town of Badme, began in 1998 and ended in what year?
2000

54. Mount Catherine is the highest point in what country bordering Libya?
Egypt

55. Southern National Park is located in what landlocked country bordering the Central African Republic?
South Sudan

56. The Matandu River, located in Tanzania, empties out into what body of water?
Indian Ocean

57. Taulud Island is a site in what country bordering the Red Sea and whose highest point is Emba Soira Mountain?
Eritrea

58. Samburu National Reserve is located in what country bordering Lake Victoria?
Kenya

59. Zanzibar Island, Pemba Island, and Mafia Island belong to what country where Lake Manyara National Park and Serengeti National Park can be found?
Tanzania

60. Blue Nile Falls is a tourist attraction southwest of Lake Tana and is located in what hugely-populated country where Lake Zway and Simien Mountains National Park can be found?
Ethiopia

Southern Africa
South Africa, Madagascar, Zimbabwe, Angola, Botswana, Mozambique, Namibia, Zambia, Malawi, Mauritius, Comoros, Seychelles, Reunion, Lesotho, Swaziland

1. The Huila Plateau is located in what country bordering Bengo Bay and Zambia?
Angola

2. People have been using Ngywenya Mine for 43,000 years, and that makes it the oldest mine in the world.

This mine is located in what country where the Orange River forms part of the country's border with Namibia?
South Africa

3. Antongil Bay borders what country famous for its Tsingy de Bemaraha National Park?
Madagascar

4. Pamplemousses is a city in what country?
Mauritius

5. The Glorioso Islands, located within a marine protected area, are situated in the Indian Ocean and belong to what country?
France

6. Karthala is a famous volcano in what country northwest of Madagascar?
Comoros

7. Lake Kariba borders Zambia and what other country containing the Shangani and Mwenezi Rivers?
Zimbabwe

8. Hwange National Park is located in what country bordering South Africa?
Zimbabwe

9. The Vaal River, whose source in in the Drakensberg Mountains, is a tributary of what river?
Orange River

10. Skeleton Coast is a region in what country?
Namibia

11. The Rempart River is in what country whose capital is Port Louis?
Mauritius

12. The junction of the Orange River and Makhaleng River is the lowest point in what greatly elevated country bordering South Africa?
Lesotho

13. Zakouma National Park is located in what country?
Chad

14. Central Kalahari Game Reserve is located in what country with the Okavango Delta?
Botswana

15. Schuckmannsburg is west of the Zambezi River in what country bordering Hottentot Bay?
Namibia

16. Lake Ngami is located in what country home to Lake Xau and the Makgadikgadi Pans?
Botswana

17. The Cahora Bassa Dam was built across the Zambezi River in 1974 in what country?
Mozambique

18. Lilongwe is the capital of what country whose major cities include Mzuzu, Zomba, and Blantyre?
Malawi

19. Victoria Falls is on what country's border with Zambia?
Zimbabwe

20. The Muchinga Mountains are miles north of Lower Zambezi National Park in what country bordering Angola?
Zambia

21. Juan de Nova Island, situated in the Mozambique Channel and west of Madagascar, belongs to what country?
France

22. Mamoudzou is a city that is the capital of what French territory in the Indian Ocean?
Mayotte

23. Rodrigues is an island belonging to what country where Hindi, Bhojpuri, and French are spoken?
Mauritius

24. The islands of Agalega are two Outer Islands belonging to what country northeast of Reunion?
Mauritius

25. Augrabies Falls is located in what country bordering St Helena Bay?
South Africa

26. Lobamba is the legislative capital of what country?

Swaziland

27. Boulders Beach is on the Cape of Good Hope in what country?
South Africa

28. The Matopos Hills are located in what country home to the major cities of Chitungwiza and Harare?
Zimbabwe

29. Cabinda is an exclave of what country bordering the Democratic Republic of the Congo?
Angola

30. Tofo Beach is located in what country containing the Zambezi River Delta and Cape Sao Sebastiao?
Mozambique

31. Monte Binga is a peak in what country whose capital is Maputo?
Mozambique

32. Lake Bangweulu is located in what country bordering the Caprivi Strip?
Zambia

33. Busanga Swamp is located in what landlocked country?
Zambia

34. Barra Falsa Point and Barra Point are in what country bordering Tanzania?
Mozambique

35. The Omatako River is in what country containing Kaokoland?
Namibia

36. Rodrigues is the second largest island in what country whose capital is Port Louis?
Mauritius

37. Bengo Bay is north of what city that is the capital of Angola?
Luanda

38. The Huila Plateau is south of the Bie Plateau in what country?
Angola

39. Kasanka National Park is west of the Muchinga Mountains in what country?
Zambia

40. Matusadona National Park is located south of what lake in Zimbabwe?
Lake Kariba

41. Central Kalahari Game Reserve is located in the central region of what landlocked country?
Botswana

42. Lake Ngami is located west of Toteng and southwest of Maun, which are cities in what country?
Botswana

43. Great Namaland is in the southern region of what country bordering South Africa?
Namibia

44. Hottentot Bay, in the Atlantic Ocean, borders what country?
Namibia

45. Boulders Beach is located on the Cape of Good Hope in what country whose administrative capital is Pretoria?
South Africa

46. Tsingy de Bemaraha National Park is home to a maze of stone towers, canyons, and caves as well as unusual plants and animals. This national park is located in what country?
Madagascar

47. Karthala, one of the world's largest active volcanoes, is located in what country?
Comoros

48. Cousin Island is a center of rare birds in what country whose capital is Victoria?
Seychelles

49. Reunion, southwest of Mauritius, is a territory of what country in Europe?
France

50. Bhojpuri and Hindi are spoken languages in what country home to the city of Pamplemousses?

Mauritius

51. Bengo Bay is in the Atlantic Ocean on what country's coast?
Angola

52. The Huila Plateau is located west of the Cubango River in what country?
Angola

53. The Caprivi Strip borders Angola, Botswana, and what country to the north?
Zambia

54. Lake Nyasa is also known by what name?
Lake Malawi

55. Seweweekspoortpiek is a peak in what country bordering Richards Bay?
South Africa

56. Bazaruto Island belongs to what country containing the Zambezi River Delta?
Mozambique

57. Nosy Boraha, also known as Ile Saint Marie, is located south of what bay bordering Madagascar and the Masoala Peninsula?
Antongil Bay

58. Epupa Falls is located on what river?
Cunene River

59. Tiger Bay borders Marca Point in what country?
Angola

60. The Etosha Pan is located in what country containing the Caprivi Strip and Namib Desert?
Namibia

61. The Muchinga Mountains are located in what country containing part of the Zambezi River and bordering Lake Tanganyika?
Zambia

62. Mahe Island is in what island country off the eastern coast of Africa?
Seychelles

63. The ruby red fody bird is native to what country whose major language is Malagasy?
Madagascar

64. KwaZulu-Natal is a province in what country whose judicial capital is Bloemfontein?
South Africa

65. Bassas da India is a territory of what country in Europe?
France

66. The Save River has its mouth in what channel?
Mozambique Channel

67. The junction of the Orange and Makhaleng Rivers is the lowest point in what country containing the Senqunyane River?
Lesotho

68. Moroni, the capital of Comoros, is located on what island?
Grande Comore Island

69. Cape Agulhas, the southernmost point in Africa, belongs to what country?
South Africa

70. Lake Chilwa borders what thin country to the west that shares its name with a lake?
Malawi

71. Sapitwa is the highest point at 9,849 feet in what country bordering Mozambique?
Malawi

72. Banhine National Park and Zinave National Park are located in what country bordering the Indian Ocean?
Mozambique

73. Malagasy is the official language of what island country in the Indian Ocean bordering the Mozambique Channel?
Madagascar

74. Mamoudzou is the capital of what island territory of France in the Indian Ocean?
Mayotte

75. Lobito and Malanje are major cities in what country bordering Namibia?
Angola

76. Cape Fria is located in what country containing the Etosha Pan?
Namibia

77. Ovamboland and Damaraland are regions in what country containing the peak of Brandberg?
Namibia

78. Polokwane is north of what city also known as Tshwane?
Pretoria

79. The Mozambique Channel borders Mozambique and what large country to the west whose highest point is Maromokotro and official languages are Malagasy and French?
Madagascar

80. Port Louis is the largest city by population in what country where the majority of the people follow Hinduism?
Mauritius

Central Africa
Democratic Republic of the Congo, Congo, Gabon, Central African Republic, Cameroon, Equatorial Guinea, Sao Tome and Principe, Chad

1. Malabo, the capital of Equatorial Guinea, is located on what island?
Bioko

2. Sao Tome and Principe is northeast of what island belonging to Equatorial Guinea?
Annobon

3. The Ogooue River is located in what country containing Ivindo National Park?
Gabon

4. The Lagdo Reservoir is located in what country whose major cities include Douala and Yaounde?
Cameroon

5. Noubale-Ndoki National Park is located in what country whose capital is Brazzaville?
Congo

6. The Tshuapa and Aruwimi Rivers are located in what country whose major cities include Lubumbashi and Mbuji-Mayi?
Democratic Republic of the Congo

7. Dzanga-Sangha Special Reserve is a rainforest reserve in what country bordering Cameroon to the west?
Central African Republic

8. Spanish Guinea was the former name of what country whose predominant languages are Spanish, French, and pidgin English?
Equatorial Guinea

9. The Ituri Forest is located in what country containing the Congo and Lomami Rivers?
Democratic Republic of the Congo

10. Virunga National Park is a site in what country bordering the Central African Republic?
Democratic Republic of the Congo

11. The goliath frog, the largest frog in the world, lives in the rainforests of Equatorial Guinea and what country containing part of the Benue River?
Cameroon

12. Cape Lopez and Pongara Point are located in what country whose capital is Libreville?
Gabon

13. The island of Bioko in Equatorial Guinea is located in what gulf?
Gulf of Guinea

14. The Okapi Wildlife Reserve is located in what country bordering Lake Tanganyika?
Democratic Republic of the Congo

15. Emi Koussi is located in what mountain range in Chad?
Tibesti Mountains

16. Lake Mweru is located on the border between Zambia and what country containing the Katanga Plateau?
Democratic Republic of the Congo

17. Kisangani is a city in what country containing the Kasai River?
Democratic Republic of the Congo

18. Nki National Park is located in the southeastern region of what country?
Cameroon

19. What river in Africa is the second largest in the world by discharge?
Congo River

20. The manatee is the symbol of Conkouati-Douli National Park. This park is in what country bordering the Democratic Republic of the Congo?
Congo

21. The Tibesti Mountains are in the northern part of what country containing the Aozou Strip?
Chad

22. The Mitumba Mountains are located in what country containing the Sankuru and Salonga Rivers?
Democratic Republic of the Congo

23. The Kwango River is a tributary of what river that is a tributary of the Congo River, the tenth largest river in the world?
Kwa River

24. The Cuango River forms part of Angola's border with what country bordering Cabinda?
Democratic Republic of the Congo

25. Sinianka-Minia Game Reserve is in what country bordering Cameroon to the southwest?
Chad

26. Oyala is a city being built to become the future capital of what country, replacing Malabo?
Equatorial Guinea

27. The Bakassi Peninsula is governed by what country bordering Nigeria to the west?
Cameroon

28. Emi Koussi is the highest mountain in what desert?
Sahara Desert

29. Ivindo National Park is located in the eastern region of what country?
Gabon

30. Pongara Point is located in what country whose capital is Libreville?
Gabon

31. Waza National Park is located in the northern region of what country bordering the Central African Republic?
Cameroon

32. The Lagdo Reservoir is located miles north of the Mbang Mountains in what country?
Cameroon

33. Port-Gentil is a city on Cape Lopez in what country?
Gabon

34. Pico de Sao Tome is a volcanic mountain covered with rainforests in what country?
Sao Tome and Principe

35. Cabinda, an exclave of Angola, is located south of the Congo and west of what country?
Democratic Republic of the Congo

36. Loango National Park is located in what country bordering the Atlantic Ocean to the west?
Gabon

37. Sarh, Moundou, and Abeche are major cities in what country?
Chad

38. Bioko borders what bight to the northwest?
Bight of Bonny

39. The Nyong River flows into what major gulf?
Gulf of Guinea

40. Fitri Lake is located in what country bordering Cameroon to the southwest?
Chad

41. Boyoma Falls is located in the northern region of what country?
Democratic Republic of the Congo

42. The Mitumba Mountains are located in the southwestern region of what country?
Democratic Republic of the Congo

43. The Mbakaou Reservoir is located in what country bordering Equatorial Guinea?
Cameroon

44. Gabon, Angola, Cameroon, the Democratic Republic of the Congo, and the Central African Republic border what country?
Congo

45. Zakouma National Park and the Ergig River can be found in what country?
Chad

46. Franceville and Tchibanga are major cities in what country?
Gabon

47. Ureca is a city on the island of Bioko in what country?
Equatorial Guinea

48. Lake Kivu and Lake Albert border what country to the west?
Democratic Republic of the Congo

49. Conkouati-Douli National Park is located in what country bordering the Central African Republic and Cameroon?
Congo

50. Mbuji-Mayi, Lubumbashi, and Kinshasa are major cities in what country bordering Lang Tanganyika?
Democratic Republic of the Congo

North Africa
Algeria, Tunisia, Libya, Morocco, Western Sahara

1. The Gulf of Sidra, part of the Mediterranean Sea, borders what country?
Libya

2. Tunisia borders what gulf to the east that is part of the Mediterranean Sea?
Gulf of Gabes

3. The Monument Des Martyrs honors soldiers that fought in what country's war for independence?
Algeria

4. One of the world's largest amphitheaters in history is located in El Jemm in what country?
Tunisia

5. The Libyan Plateau is located in Libya and what other country?
Egypt

6. The Jebel Acacus Mountain Range is located in the southwestern corner of what country?
Libya

7. The Hauts Plateaux, extending into Algeria, is located in what country whose lowest point is Sebkha Tah?
Morocco

8. The Qerqenah Islands belong to what country separated from Sicily by the Strait of Sicily?
Tunisia

9. The Gulf of Hamamet forms the northeastern coast of what country?
Tunisia

10. Tamanrasset is a city in what country bordering the Alboran Sea?
Algeria

11. The Atlas Mountains are in Morocco, Tunisia, and what country?
Algeria

12. Libya borders what gulf to the north?
    Gulf of Sidra

13. Leptis Magna was a city belonging to an empire. The ruins of this city is located in Libya, and it belonged to what empire thousands of years ago?
    Roman Empire

14. The Ez-Zitouna Mosque is a major mosque in what country?
    Tunisia

15. The Great Eastern Erg and Great Western Erg are located in what country containing the Ahaggar Mountains?
    Algeria

16. The Gulf of Bomba is north of what country whose capital is Tripoli?
    Libya

17. The Gurgi Mosque in Tripoli is located in what country?
    Libya

18. Sabkhat Ghuzayyil is the lowest point in what predominantly Muslim country bordering Tunisia?
    Libya

19. Fez is one of the major cities in what country bordering Western Sahara?
    Morocco

20. Libya and Tunisia border what major sea to the north?
    Mediterranean Sea

21. Sabha is a major city in what country bordering the Gulf of Sidra?
Libya

22. Cyrenaica and Fezzan are regions of what country bordering Algeria to the west?
Libya

23. Bechar and Setif are cities in what country that is the largest in Africa?
Algeria

24. The Tropic of Cancer passes through Egypt, Western Sahara, Mauritania, Niger, Mali, Libya, and what other country in Africa?
Algeria

25. Sfax is a major city in what small country?
Tunisia

26. Jebel Toukbal is a peak in what country bordering the Strait of Gibraltar?
Morocco

27. Ceuta and Melilla in Morocco belong to what European country?
Spain

28. What mountain is the highest in Tunisia?
Jebel ech Chambi

29. The Hauts Plateaux can be found in Algeria and what country bordering the Atlantic Ocean?
Morocco

30. The Ahaggar Mountains are in the southern region of what country that is the largest in Africa by area?
Algeria

31. Mount Tahat, in the Ahaggar Mountains, is the highest point in what country?
Algeria

32. Tripoli is a major city and the capital of what country bordering the Mediterranean Sea to its north?
Libya

33. Misratah is a port city on the Mediterranean Sea in what country?
Libya

34. What is the largest city in the Moroccan territory of Western Sahara?
Laayoune

35. Saguia el Hamra is a region in what territory whose capital is Laayoune?
Western Sahara

36. What is the largest religion followed in Libya?
Islam

37. Ras Cantin and Cap Juby are located in what country?

Morocco

38. Western Sahara borders what country to the north?
Morocco

39. Sidi Salem Dam is located on the Medjerda River in what country?
Tunisia

40. The Amour and Ksour Mountain Ranges are located in what country bordering the Mediterranean Sea to the north?
Algeria

West Africa
Mauritania, Mali, Nigeria, Ghana, Ivory Coast, Guinea-Bissau, Gambia, Senegal, Guinea, Sierra Leone, Liberia, Burkina Faso, Togo, Benin, Cape Verde

1. French Sudan and Sudanese Republic were former names of what country whose highest point is Hombori Tondo at 3,789 feet?
Mali

2. The Senegal River forms the entire border between Mauritania and what country?
Senegal

3. Fouta Djallon is a highland region south of the Sahel in what country bordering the Atlantic Ocean?
Guinea

4. Yamoussoukro is the legislative capital of what country?
Ivory Coast

5. "W" National Park is located in the southwestern region of what country whose capital is Niamey?
Niger

6. Lake Volta is located in what country containing Kakum National Park and Akosombo Dam?
Ghana

7. The Kainji Reservoir is formed by the Kainji Dam in what country?
Nigeria

8. The Cacheu River is located in what country whose capital is Bissau?
Guinea-Bissau

9. Santo Antao and Sao Nicolau are islands in what archipelagic country?
Cape Verde

10. Kachikally Crocodile Pool is located in Serekunda in what country?
Gambia

11. The Saloum and Sine Rivers are located in what country whose capital is Dakar?
Senegal

12. The Nimba Mountains are located in Guinea, Ivory Coast, and what other country?
Liberia

13. Orango Islands National Park is located on the island of Orango in what country?
Guinea-Bissau

14. Bossou Forest is located in the southern region of what country where the Niger and Milo Rivers can be found?
Guinea

15. The Stone Sculptures of Laongo are located northeast of Ouagadougou in what country?
Burkina Faso

16. Lake Volta is the largest lake in what country whose major cities include Kumasi and Tamale?
Ghana

17. Cotonou is the Seat of Government in what country bordering Togo?
Benin

18. Kakum National Park is located in what country containing Cape Three Points?
Ghana

19. Akloa Falls is located in Badou in what country?
    Togo

20. The Wechiau Community Hippo Sanctuary is located in Wechiau in what country?
    Ghana

21. The Ouidah Museum of History is located in what country?
    Benin

22. The Niger and Bani Rivers are located in what country containing the cities of Mopti and Sikasso?
    Mali

23. Rhoko Forest is located southeast of the cities of Onitsha and Enugu in what country famous for its oil production?
    Nigeria

24. Tarkwa Beach is one of the most popular beaches in Lagos in what country bordering the Bight of Benin and the Bight of Bonny?
    Nigeria

25. The Bokel and Moa Rivers are located in what country bordering Yawri Bay?
    Sierra Leone

26. The Kaduna and Gongola Rivers are in what country?
    Nigeria

27. The Turner's Peninsula is in what country?
    Sierra Leone

28. The Bissagos Islands belong to what country bordering Senegal and Guinea?
Guinea-Bissau

29. Cape Boujdour is located in what disputed area?
Western Sahara

30. Sal and Maio are islands in what country off the coast of Africa?
Cape Verde

31. The Mangueni Plateau is located in what country home to the Dillia River?
Niger

32. Lake Faguibine is located in what country bordering Mali?
Niger

33. The Bauchi Plateau is located in what country bordering the Gulf of Guinea?
Nigeria

34. The Mbang Mountains are located in what country bordering the Bight of Bonny?
Cameroon

35. Fitri Lake is located in what country with the Tibesti Mountains?
Chad

36. Mbandaka and Lubumbashi are major cities in what country that borders South Sudan and the Central African Republic?
Democratic Republic of the Congo

37. The Bight of Bonny separates Bioko from what country whose capital is Yaounde?
Cameroon

38. Ndogo Lagoon and Nkomi Lagoon are in what country along the Atlantic Coast?
Gabon

39. Livingstone Falls is on what major Central African river?
Congo River

40. Banc D'Arguin National Park is located in what country bordering Levrier Bay?
Mauritania

41. The Ouadou River is located in what country bordering Senegal to the south?
Mauritania

42. The Tree of Tenere, the last tree standing in the Sahara until 1973, is on display in what country bordering Chad?
Niger

43. The Dosso Reserve is located in Dosso in what country?
Niger

44. Yankari National Park is southwest of the cities of Gombe and Kumo in what country?
Nigeria

45. What is the official currency of Nigeria?
Naira

46. Yawri Bay borders what country whose capital is Freetown?
Sierra Leone

47. The Mano and Loffa Rivers are in what country bordering Sierra Leone, Guinea, and Ivory Coast?
Liberia

48. Olumo Rock is a sacred place in what country containing Yankari National Park?
Nigeria

49. The Iguidi Desert in Algeria extends into what country whose capital is Nouakchott?
Mauritania

50. Agadez and Zinder are cities in what country bordering Burkina Faso and Mali?
Niger

51. The Jos Plateau is located in what country bordering the Gulf of Guinea?
Nigeria

52. The Air Mountains are located in what country that shares the same name as a river?
Niger

53. Monrovia is a city in the Montserrado County of what country?
Liberia

54. The Plateau of Djado is a desert basin in what desert?
Sahara Desert

55. The Yamoussoukro Basilica is a church that might be the largest Christian house of worship in the world. It is in what country bordering the Gulf of Guinea?
Ivory Coast

56. Cha Das Caldeiras is a village inside the crater of a volcano on Fogo Island in what country?
Cape Verde

57. What country, bordering Niger and Benin, is Africa's largest producer and exporter of oil?
Nigeria

58. Adamawa and Kano are states in what country bordering the Gulf of Guinea?
Nigeria

59. Boucle du Baoule National Park is in what landlocked country bordering Niger and Burkina Faso?
Mali

60. Chappal Waddi is the highest point at 7,936 feet in what country?
Nigeria

61. Mount Wuteve is the highest point in what country whose capital city is a chief port city on the Atlantic Ocean?
Liberia

62. What is the predominant European language spoken in Togo?
French

63. Kediet ej Jill, at 915 meters, is the highest point in what country?
Mauritania

64. The Macina Swamp can be found in what country?
Mali

65. Sebkha de Ndrhamcha, a salt pan, is the lowest point in what country whose capital is Nouakchott?
Mauritania

66. The Saloum River is located completely in what country?
Senegal

67. What city is the administrative capital of Cote d'Ivoire?
Abidjan

A Competitor's Compendium to the Geography Bee

Australia and Oceania

Australia
Australia

1. The Great Sandy Desert is located in what Australian state whose capital is Perth?
   Western Australia

2. Boomerang Beach is a famous place in what state bordering Queensland to the north?
   New South Wales

3. The Bass Strait separates what state from Victoria?
   Tasmania

4. Lake Eyre is located in what state occupying Kangaroo Island and bordering Lacepede Bay?
   South Australia

5. The Great Barrier Reef borders what state to the west?
   Queensland

6. The Tropic of Capricorn passes through Western Australia, Queensland, and what federal territory?
   Northern Territory

A Competitor's Compendium to the Geography Bee

7. The Great Australian Bight is south of what two states?
   Western Australia and South Australia

8. Cape Naturaliste is located in what state bordering the Joseph Bonaparte Gulf?
   Western Australia

9. Sydney and Canberra are located in what state west of South Australia?
   New South Wales

10. Hobart is the largest city in what state?
    Tasmania

11. Lake Argyle is primarily located in what state?
    Western Australia

12. Melville Island is off the coast of what federal territory?
    Northern Territory

13. Brisbane is the largest city in what state that contains part of the Great Artesian Basin?
    Queensland

14. The Simpson Desert is in the southwestern region of Queensland and the southeastern region of what federal territory?
    Northern Territory

15. Melbourne is located in what small state bordering the Tasman Sea?
    Victoria

16. The Murray-Darling River System is located in what region of Australia – East or West?
    East

17. Queensland borders what major gulf to the northwest?
    Gulf of Carpentaria

18. Cape Melville and Cape York are located in what state?
    Queensland

19. The Sydney Opera House, one of the most photographed buildings in the world, is located in what state?
    New South Wales

20. The Nullarbor Plain and the Great Victoria Desert are located in Western Australia and what other state?
    South Australia

21. Lake Torrens is located in what state bordering Spencer Gulf?
    South Australia

22. The Victoria and Daly Rivers are located in what federal territory containing part of Lake Mackay?
    Northern Territory

23. Fraser Island is off the coast of what state that is the second largest in Australia?
    Queensland

24. The Flinders River is in what state containing Brisbane and Rockhampton?
Queensland

25. Melbourne is the largest city in terms of population in what state bordering New South Wales?
Victoria

26. Lake Barlee is located in what country bordering Roebuck Bay and the Great Australian Bight?
Western Australia

27. Cape Leeuwin is located in what state bordering the Indian Ocean?
Western Australia

28. The Northern Territory borders what sea directly to the north that is northeast of the Timor Sea?
Arafura Sea

29. The Lachlan River is in what country containing the cities of Sydney and Canberra?
New South Wales

30. The Bass Strait separates Tasmania from what state to the north?
Victoria

31. Witjira National Park and Walakara Indigenous Protected Area are located in what state containing the Musgrave Ranges and the Strzelecki Desert?
South Australia

32. Cape Naturaliste in Western Australia borders what bay?
Geographe Bay

33. The Great Barrier Reef is located northeast of what Australian state?
Queensland

34. Mount Gambier is the second most populous city in what Australian state?
South Australia

35. The island of New Guinea is separated from Australia by what sea?
Arafura Sea

36. The Gibson Desert is in what state in Australia?
Western Australia

37. Ipswich and Cairns are cities in what Australian state bordering South Australia?
Queensland

38. Lacepede Bay borders what state to the north where the major city of Adelaide is located?
South Australia

39. Mount Kosciuzko is the highest point in Australia, and is located in what state ordering Queensland?
New South Wales

40. The Cobourg Peninsula and the Wessel Islands are located in what federal territory?
Northern Territory

Melanesia and Papua New Guinea
Fiji, Papua New Guinea, Vanuatu, Solomon Islands, New Caledonia

1. Varirata National Park is located in what country?
Papua New Guinea

2. The Louisiade Archipelago belongs to what country whose capital is Port Moresby?
Papua New Guinea

3. New Britain and New Ireland are islands belonging to what country that borders the Solomon Sea?
Papua New Guinea

4. The Sri Siva Subramaniya Temple, the largest temple in the Southern Hemisphere, is located in Nadi in what country that is mainly Christian and Hindu?
Fiji

5. Honiara is the capital of what country?
Solomon Islands

6. Noumea is the capital of what territory of France miles south of Vanuatu?
New Caledonia

7. The Lau Island Group belongs to what country containing the Garden of the Sleeping Giant?
Fiji

8. Malakula is the second largest island in what country whose capital is Port Vila?
Vanuatu

9. Rotuma belongs to what country containing the Yasawa Island Group?
Fiji

10. New Georgia Sound borders what country whose largest island is Guadalcanal?
Solomon Islands

11. The Bismarck Archipelago borders what sea?
Bismarck Sea

12. Mt. Wilhelm is a peak in what country containing the Kikori and Strickland Rivers?
Papua New Guinea

13. Lake Murray is located in what country bordering the Ysabel Channel and Wide Bay?
Papua New Guinea

14. Tanna Island is located in what country?

Vanuatu

15. The Vatuira Channel is located in what country?
    Fiji

16. Ambrym Volcano is one of the most active volcanoes in Oceania, located in what country?
    Vanuatu

17. Espiritu Santo is the largest island in what Melanesian archipelagic country?
    Vanuatu

18. Vanua Levu and Viti Levu are islands in what country in the Pacific Ocean?
    Fiji

19. Tok Pisin is a creole language spoken in what country?
    Papua New Guinea

20. The Louisiade Archipelago is south of what sea?
    Solomon Sea

21. The Santa Cruz Islands are part of what country?
    Solomon Islands

22. Malakula and Erromango are islands in what country whose capital is Port Vila?
    Vanuatu

23. The Yasawa Group is located in what country home to the Sri Siva Subramaniya Temple?

Fiji

24. Choiseul and Santa Isabel are islands in what country?
Solomon Islands

25. Kadavu and Vatoa are islands belonging to what country?
Fiji

26. Kimbe Bay borders what island in Papua New Guinea to the south?
New Britain

27. The Strickland and Ramu Rivers are located in what country?
Papua New Guinea

28. The Loyalty Islands belong to what French territory?
New Caledonia

29. Espiritu Santo is the largest island by area in what country?
Vanuatu

30. The Lau Group is an archipelago of islands belonging to what country whose second largest island is Vanua Levu?
Fiji

31. Rennell is an island belonging to what country bordering the New Georgia Sound?
Solomon Islands

32. The Banks Islands are an archipelago belonging to what country?
Vanuatu

33. Varirata National Park is located in the southeastern region of what country?
Papua New Guinea

34. Jacquinot Bay and Rabaul are cities on the island of New Britain in what country bordering the Bismarck Archipelago?
Papua New Guinea

35. The Ninigo Group and the Hermit Islands belong to what country bordering the Coral Sea and the Goschen Strait?
Papua New Guinea

36. What is the highest point in Fiji, at 4,344 feet?
Tomanivi

37. The vatu is the official currency of what country hundreds of miles west of Fiji?
Vanuatu

38. Mount Popomanaseu is the highest point in what country?
Solomon Islands

39. The Shepherd Islands and the Torres Islands belong to what country?
Vanuatu

A Competitor's Compendium to the Geography Bee

40. Suva, on the island of Viti Levu, is the most populous and largest city in what country?
Suva

Polynesia and New Zealand
New Zealand, Samoa, Tuvalu, Tonga, American Samoa, French Polynesia, Cook Islands, Niue, Tokelau, Midway Islands, Pitcairn Islands

1. The Southern Alps are located in what country bordering Pegasus Bay?
New Zealand

2. The Coromandel Peninsula is on what island?
New Zealand

3. Auckland is the largest city in what country?
New Zealand

4. The Marquesas Islands, north of the Tuamotu Archipelago, belong to what country?
France

5. Niue is a territory of what country?
New Zealand

6. Nuku'alofa is the capital of what country containing Tongatapu Island?

Tonga

7. The Alofaaga Blowholes are located in what country miles south of Tokelau?
Samoa

8. Honolulu is the capital of what U.S. state geographically located in Polynesia?
Hawaii

9. Funafuti is the capital of what country miles north of Fiji?
Tuvalu

10. Milford Sound is a famous tourist attraction on what island containing the cities of Queenstown and Christchurch?
South Island

11. Manukau, also known as Manukau Central, is southwest of what major New Zealand city?
Auckland

12. The Cook Islands belong to what country?
New Zealand

13. Samoa is located west of American Samoa, a territory of the United States. What is the capital of Samoa?
Apia

14. Golden Bay and Tasman Bay border what island containing Cape Farewell and the Banks Peninsula?
New Zealand

15. The Maori are the native people of what country home to the cities of Tauranga and Whangarei?
New Zealand

16. Mount Ruapehu is located in what mountain range in New Zealand?
Southern Alps

17. The Shotover River is a tourist attraction in what country containing North Island and South Island?
New Zealand

18. Mitre Peak rises more than 5,550 feet over Milford Sound in what country?
New Zealand

19. The Ellice Islands was the original name of what country?
Tuvalu

20. Nuku'alofa is a city in Tonga on what island that is the largest in the country?
Tongatapu

21. Franz Josef Glacier is located in Westland Tai Poutini National Park in what country?
New Zealand

22. Tatakoto and Pukapuka are atolls located in what major archipelago?
Tuamotu Archipelago

23. What city is the capital of French Polynesia?
Papeete

24. Palmerston Atoll and Aitutaki Atoll are located in the Cook Islands. This island group is a territory of what country?
New Zealand

25. Swains Island is located in what U.S. territory east of Samoa and south of Tokelau?
American Samoa

26. Tokelau is a territory of what country?
New Zealand

27. Lake Ellesmere and Lake Taupo are located in what country?
New Zealand

28. What is the capital of American Samoa and is also the country's most populous city?
Pago Pago

29. Taumatawhakatangihangakoauauotamateapokaiwhenakitanatahu is a peak on North Island in what country?
New Zealand

30. Milford Track is a trail on South Island in what country?
New Zealand

A Competitor's Compendium to the Geography Bee

Micronesia
Federated States of Micronesia, Palau, Nauru, Marshall Islands, Kiribati, Northern Mariana Islands, Guam, Jarvis Island, Palmyra Atoll, Howland Island, Baker Island

1. Guam is south of what U.S. Territory whose capital is Saipan?
Northern Mariana Islands

2. The Caroline Islands are part of what country whose capital is Pohnpei?
Federated States of Micronesia

3. Palmyra Atoll, Howland Island, Baker Island, and Jarvis Island are close to what country?
Kiribati

4. Melekeok is the capital of what country west of the Caroline Islands?
Palau

5. The Ngardmau Waterfalls are located in what country?
Palau

6. The Alele Museum is located in what country bordering the Ratak Chain and Ralik Chain?
Marshall Islands

7. Majuro is the capital of what island group?
   Marshall Islands

8. Nan Madol is an ancient site in what city?
   Pohnpei

9. Babelthuap is the largest island in what country?
   Palau

10. Guam is south of what territory of the United States?
    Northern Mariana Islands

11. Buada Lagoon is located in what country bordering Anibare Bay to the east?
    Nauru

12. Meneng Point is located in what country with a population of approximately ten thousand people?
    Nauru

13. Saipan is the capital of what U.S. territory?
    Northern Mariana Islands

14. The Mariculture Demonstration Center is located on Malakal Island in what country?
    Palau

15. What country is the most populous in Micronesia?
    Federated States of Micronesia

16. Rongelap Atoll and Taongi Atoll are located in what country?
Marshall Islands

17. Tabuaeran, also known as Fanning Island, belongs to what country?
Kiribati

18. Majuro is the capital of what country northwest of Kiribati?
Marshall Islands

19. Tarawa is the capital of what country?
Kiribati

20. The Phoenix Islands and the Gilbert Islands are archipelagos in what country?
Kiribati

Antarctica

1. Filchner Ice Shelf borders what large island?
   Berkner Island

2. Roosevelt Island is situated in what shelf?
   Ross Ice Shelf

3. The Polar Plateau borders what major Antarctic mountain range?
   Transantarctic Mountains

4. The American Highland borders what ice shelf?
   Amery Ice Shelf

5. What ice shelf is directly north of Queen Maud Land?
   Fimbul Ice Shelf

6. Palmer Land is located on what peninsula?
   Antarctic Peninsula

7. Maitri is a research station on Antarctica belonging to what country?
   India

8. The Latady Mountains border what mountain range to the west?
   Sweeney Mountains

9. Vincennes Bay is located in what ocean surrounding Antarctica?
Southern Ocean

10. Are there any permanent residents on Antarctica?
No

11. The Antarctic Peninsula borders what sea to the east, northeast of the Ronne Ice Shelf?
Weddell Sea

12. The Lambert Glacier is south of what ice shelf bordering the American Highland?
Amery Ice Shelf

13. Burke Island is located off the coast of Antarctica in what sea?
Amundsen Sea

14. The Bransfield Strait borders what major peninsula in Antarctica?
Antarctic Peninsula

15. The Davis Sea is west of what major ice shelf in Antarctica?
Shackleton Ice Shelf

16. The Moscow University Ice Shelf has a similar name to the capital of what country?
Russia

17. How many species of penguin live on Antarctica?

Five

18. The Antarctic Treaty was signed by 12 countries in what year?
    1959

19. How many countries lay claim over Antarctica?
    Seven

20. The Antarctic Peninsula borders what major sea to the west?
    Bellingshausen Sea

A Competitor's Compendium to the Geography Bee

Mock Bee

Classroom/School Competition:

1. Which city is the capital of Egypt – Cairo or Alexandria?
   Cairo

2. Ontario and Quebec are provinces in which country – Mexico or Canada?
   Canada

3. Argentina borders which ocean to the east – Atlantic Ocean or Pacific Ocean?
   Atlantic Ocean

4. Which country is the largest in Africa and the tenth largest in the world – Sudan or Algeria?
   Algeria

5. Which South Asian country is landlocked – Bangladesh or Bhutan?
   Bhutan

6. Tokyo, the most populous metropolitan city in the world, is located in which country – Japan or South Korea?
   Japan

7. The Falkland Islands are an overseas territory of which country in Western Europe – United Kingdom or France?
United Kingdom

8. Which Middle Eastern country is west of Afghanistan – Iran or Pakistan?
Iran

9. Spain borders France, Andorra, and which country to the west – Luxembourg or Portugal?
Portugal

10. The Panama Canal, in Panama, connects the Atlantic Ocean to which other ocean – Pacific or Indian?
Pacific

11. The islands of Sumatra and Java belong to which archipelagic country – Indonesia or Philippines?
Indonesia

12. Which country is the largest in Central Asia by area – Kazakhstan or Uzbekistan?
Kazakhstan

13. Yemen and Oman are located on which peninsula – Jaffna or Arabian?
Arabian

14. The Yucatan and Baja California peninsulas are located in which North American country – Mexico or Cuba?
Mexico

15. Tunisia borders which sea to the north – Red Sea or Mediterranean Sea?
Mediterranean Sea

16. Mount Everest is located on the border between China and which other country – India or Nepal?
Nepal

17. South Africa has how many capitals – Three or Four?
Three

18. The Maldives are located in which ocean east of Africa – Pacific Ocean or Indian Ocean?
Indian Ocean

19. The Galapagos Islands belong to which country in South America bordering Peru and Colombia – Ecuador or Brazil?
Ecuador

20. New Delhi, Chennai, Kolkata, and Mumbai, in no specific order, are four of the most populous cities in which country – India or Pakistan?
India

21. Hinduism is the majority religion in Nepal and what other South Asian country?
India

22. Tamil and Telugu are languages that are part of what language family that has speakers in India, Sri Lanka, Pakistan, Bangladesh, Nepal, Malaysia, and Singapore?

Dravidian languages

23. Machu Picchu is a 15-century Inca site in what country bordering the Pacific Ocean?
Peru

24. The Sydney Opera House is situated on Bennelong Point in what country?
Australia

25. The Persian people primarily live in what country bordering the Persian Gulf?
Iran

A Competitor's Compendium to the Geography Bee

State Qualification Test

(Ask your parents or siblings to quiz you)

1. Kyushu and Honshu are islands in what archipelagic country?
   a) China
   b) South Korea
   c) Japan
   d) Taiwan

2. Gaziantep, Bursa, and Izmir are major cities in what country bordering the Mediterranean Sea?
   a) Turkey
   b) Greece
   c) Bulgaria
   d) Lebanon

3. The Balearic Islands, located in the Balearic Sea, belong to what country bordering the Gulf of Valencia?
   a) Portugal
   b) Spain
   c) Italy
   d) Andorra

4. Guayaquil and Quito are the most populous cities in what South American country bordering the Pacific Ocean to the west?
   a) Colombia
   b) Venezuela

c) Peru
d) Ecuador

5. Tonle Sap is a lake in what country bordering Thailand and the Gulf of Thailand?
a) Cambodia
b) Laos
c) Vietnam
d) Malaysia

6. Telugu is a language spoken mainly in what state in South India?
a) Karnataka
b) Tamil Nadu
c) Kerala
d) Andhra Pradesh

7. Malabo, located in the island of Bioko, is the capital of what city in Central Africa?
a) Equatorial Guinea
b) Gabon
c) Cameroon
d) Sao Tome and Principe

8. Lake Turkana is shared by Kenya and what other country in Northeast Africa?
a) Tanzania
b) Uganda
c) Ethiopia
d) Somalia

9. Queensland and New South Wales are states in what country bordering the Joseph Bonaparte Gulf and the Coral Sea?
   a) **Australia**
   b) Papua New Guinea
   c) Fiji
   d) New Zealand

10. The Euphrates and Tigris Rivers form the Shatt al Arab, which flows into what gulf?
    a) Gulf of Oman
    b) **Persian Gulf**
    c) Gulf of Aden
    d) Gulf of Aqaba

11. The Neuschwanstein Castle is located in what country where the cities of Hamburg, Munich, and Cologne can be found?
    a) **Germany**
    b) Switzerland
    c) Austria
    d) Netherlands

12. Oulu and Espoo are major cities in what country bordering the Gulf of Bothnia?
    a) Sweden
    b) Denmark
    c) Iceland
    d) **Finland**

13. Lake Onega and Lake Ladoga are huge lakes in what transcontinental country?
a) **Russia**
b) Turkey
c) Spain
d) Kazakhstan

14. The Dzungarian Basin and the Altun Shan Mountains are located in what country?
a) Mongolia
b) **China**
c) Kyrgyzstan
d) Tajikistan

15. Budapest, the capital of Hungary, is located on what major European river?
a) Dnieper River
b) Rhone River
c) Seine River
d) **Danube River**

16. Lake Volta is the largest lake in what country where Kakum National Park and the major city of Kumasi are located?
a) Benin
b) Togo
c) **Ghana**
d) Burkina Faso

17. The Niger River Delta is located in what country bordering the Gulf of Guinea?
a) Niger
b) Cameroon
c) Equatorial Guinea
d) Nigeria

18. Brahui is a Dravidian language spoken in what country with the cities of Peshawar and Multan?
a) India
b) Bangladesh
c) Pakistan
d) Sri Lanka

19. Mindanao and Palawan are major islands in what country bordering the Luzon Strait to the north?
a) Philippines
b) Indonesia
c) Taiwan
d) Malaysia

20. Sholapur and Mysore are major cities in what country bordering the Gulf of Khambhat and the Laccadive Sea?
a) India
b) Pakistan
c) Bangladesh
d) United Arab Emirates

A Competitor's Compendium to the Geography Bee

State Competition

Round 1: The United States

1. The Kuskokwim Mountains and Chugach Mountains are located in what state?
   Alaska

2. The Pu'ukohola Heiau National Historic Site is located in what state?
   Hawaii

3. Lake Champlain is located in New York and what other state?
   Vermont

4. The Gulf of Santa Catalina borders what state to the east?
   California

5. The Coconino Plateau is located in what state whose capital is Phoenix?
   Arizona

6. Georgian Bay is located in which of the Great Lakes bordering Ontario and Michigan?
   Lake Huron

7. Delaware Bay borders Delaware and what other state situated east of Pennsylvania?
New Jersey

8. Lake Pontchartrain is located in what state bordering the Gulf of Mexico to the south and Mississippi to the east?
Louisiana

9. The Everglades and Lake Okeechobee are located in what state?
Florida

10. Fort Sumter National Monument is a site in what state whose capital is Columbia?
South Carolina

11. The Edwards Plateau and Galveston are located in what state?
Texas

12. Casper and Green River are cities in what state that is the least populous in the United States?
Wyoming

13. The geographic center of the contiguous 48 states is located in what state?
Kansas

14. Akron and Dayton are major cities in what state bordering Lake Erie to the north?
Ohio

15. Nantucket Island and Martha's Vineyard are located in what state?
Massachusetts

16. The Connecticut River, which forms the border between New Hampshire and Vermont, flows into what sound south of Connecticut?
Long Island Sound

Round 2: Cultural Geography

1. Tamil, Telugu, Kannada, and Malayalam are the four most spoken languages in what language group?
Dravidian languages

2. What is the majority religion in Russia, Canada, and Brazil?
Christianity

3. Kiswahili is one of the official languages of what country bordering Lake Tanganyika to the west and Lake Malawi to the south?
Tanzania

4. Amharic and Oromo are languages spoken in Northeast Africa in what landlocked country?
Ethiopia

5. The Shwedagon Pagoda is covered with thousands of diamonds and is a sacred Buddhist site in what country?
Myanmar

6. The Batu Caves are a religious Hindu site in what country bordering the South China Sea?
Malaysia

7. What language is the official language of Yemen and Oman?
Arabic

8. Quechua was the language of what South American empire?
Incan Empire

9. Soccer is the most popular sport in what country bordering the Rio de la Plata and the Atlantic Ocean to the south?
Uruguay

10. Catalan is the official language of what country whose highest point is Pic de Coma Pedrosa?
Andorra

11. The Maori are the indigenous people of what country bordering the Cook Strait?
New Zealand

12. The Chewa tribe can be found in Lilongwe, the capital of what African country?
Malawi

13. You can find many beautifully woven and decorated Persian carpets in what country bordering the Strait of Hormuz?
Iran

14. There are many nomadic people in what landlocked country bordering China?
Mongolia

15. Khmer is the official language of what country whose currency is the riel?
Cambodia

16. Dzongkha, the national language of Bhutan, belongs to what major language family?
Sino-Tibetan languages

Round 3: Physical Geography

1. What term describes one of the seven main landmasses on the Earth's surface?
Continent

2. An area with little or no human settlement, and either little plant growth or with forests is commonly known, specifically in Australia as what?

A Competitor's Compendium to the Geography Bee

Bush

3. What geographic term describes a group or chain of islands?
   Archipelago

4. What is the name of a map that shows a small area in great detail?
   Large-scale map

5. A very hot or very cold region that is dry and receives little precipitation is known by what name?
   Desert

6. A muddy area along tropical coasts with mangrove trees is known as what kind of swamp?
   Mangrove swamp

7. The lines east and west of the Prime Meridian are known as what?
   Longitude

8. A long, narrow strip of land made of usually silt or sand that extends into a body of water from the land is known as what?
   Spit

9. A large, slow-moving mass of ice is known by what term?
   Glacier

10. What term describes a tropical grassland, or a grassy plain with few trees?

Savannah

11. What is the term describing a dense forest found in hot and wet regions near or on the Equator?
Rainforest

12. A landmass completely surrounded by water and smaller than a continent is known by what name?
Island

13. A body of water surrounded by land is known by what name?
Lake

14. What term describes a lowland formed by gravel, sand, and silt deposited by a river at the river's mouth?
Delta

15. A deep and narrow valley created by a river having steep sides is known by what name?
Canyon

16. A steep rock face on the side of a mountain and sometimes along a coast is known as a what?
Cliff

Round 4: Economic Geography

1. What country in South America is the world's largest producer of copper?
   Chile

2. What country is the world's second largest producer of iron ore?
   Australia

3. What Middle Eastern country ranks fourth in aluminum production, behind Canada?
   United Arab Emirates

4. What country is the world's third largest producer of bauxite?
   Brazil

5. What country bordering the Mediterranean Sea is the world's largest producer of apricots?
   Turkey

6. What country ranks second in world zinc production?
   Peru

7. What country in East Asia ranks first in apple production?
   China

8. What country produces the highest amount of barley in the world?
   Russia

9. What country in South Asia is the world's third largest producer of coal?

India

10. Benghazi is a major port city in Libya on what sea?
Mediterranean Sea

11. What country on the Arabian Peninsula is the world's largest exporter of oil?
Saudi Arabia

12. Which two countries export the most cotton in the world?
United States and India

13. What country ranks first in natural gas production, ahead of Russia and the European Union?
United States

14. What country in Western Europe is the world's largest exporter of cars, ahead of Japan?
Germany

15. What country in Asia exports the most computers in the world?
China

16. What country in South Asia ranks fourth in world steel production?
India

A Competitor's Compendium to the Geography Bee

Round 5: The World

1. The Andaman and Nicobar Islands belong to what country bordering the Bay of Bengal?
   India

2. The Great Karoo and the Orange River are located in what country bordering the Indian Ocean to the east and the Atlantic to the west?
   South Africa

3. The Mekong River Delta is located in what country whose most populous city is Ho Chi Minh City?
   Vietnam

4. Chiapas and Oaxaca are states in what country bordering the Gulf of Tehuantepec?
   Mexico

5. The Volga River, the longest river in Europe, empties out into what lake?
   Caspian Sea

6. The Sea of Marmara borders what transcontinental country?
   Turkey

7. Hainan is an island belonging to what country with the major cities of Fuzhou and Nanjing?

China

8. The Tibesti Mountains are located in the northern region of what country whose capital is N'Djamena?
Chad

9. Lake Baikal is located in what country bordering the Laptev Sea?
Russia

10. Kaunas and Klaipeda are cities in what country bordering Latvia to the north?
Lithuania

11. Meru National Park and Tsavo East National Park are located in what country bordering Ungama Bay, Somalia to the east, and Ethiopia to the north?
Kenya

12. Tasmania is a state and an island in what country?
Australia

13. The Baikonur Cosmodrome, a Russian-administered site, is located in what country?
Kazakhstan

14. The Troodos Mountains are located on what island country in the Mediterranean Sea whose capital is Nicosia?
Cyprus

15. Andorra is a landlocked country bordering France and what other country?
    Spain

16. Fyn and Sjaelland are islands in what country bordering the Kattegat and the Skagerrak?
    Denmark

Round 6: Political Geography

1. What island sovereign state, part of China, is trying to maintain that there are two political entities?
   Taiwan

2. Azerbaijan claims what disputed region whose capital is Stepanakert?
   Nagorno-Karabakh

3. South Ossetia is a disputed region bordering Russia to the north and what other country?
   Georgia

4. Abkhazia is a disputed region recognized by five UN member states in what country?
   Georgia

5. The Falkland Islands in South America is a territory of what country in Western Europe?
United Kingdom

6. The Faroe Islands are administered by what Scandinavian country?
Denmark

7. Christmas Island, hundreds of miles south of Java, an island in Indonesia, is a territory of what country?
Australia

8. What sovereign state is Denmark's largest territory by area?
Greenland

9. The Aland Islands are a territory of what Nordic country to the east?
Finland

10. Svalbard, whose capital is Longyearbyen, is an archipelagic territory of what country?
Norway

11. The Cook Islands, in the Pacific Ocean, belong to what country?
New Zealand

12. What sovereign state in Somalia, whose capital is Hargeysa, declares itself an independent republic but is not internationally recognized?
Somaliland

13. Kashmir is a disputed region between India and what country to the west?
    Pakistan

14. China claims land in much of what state in India?
    Arunachal Pradesh

15. Martinique and New Caledonia are territories of what country?
    France

16. Palmyra Atoll and Wake Island are territories of what country?
    United States

Round 7: U.S. National Parks

1. Death Valley National Park is located in what state with the cities of Oakland and Sacramento?
   California

2. Badlands National Park is located in the western region of what country bordering Minnesota to the east?
   South Dakota

3. Mammoth Cave National Park is located in what landlocked state whose capital is Frankfort?
Kentucky

4. Haleakala National Park is located on what island in Hawaii?
Maui

5. Grand Canyon National Park is located in what country bordering the Mexican state of Sonora to the south?
Arizona

6. Everglades National Park is located in what state bordering the Gulf of Mexico?
Florida

7. Shenandoah National Park is located in what state bordering Chesapeake Bay?
Virginia

8. Denali National Park is located in what state bordering British Columbia?
Alaska

9. Crater Lake National Park is located in what state bordering the Pacific Ocean to the west?
Oregon

10. Yellowstone National Park is located in the northwest region of what landlocked state?
Wyoming

11. Rocky Mountain National Park is located miles southwest of Fort Collins in what state?
Colorado

12. Olympic National Park is south of the Strait of Juan de Fuca in what state?
Washington

13. Zion National Park is west of Grand Staircase Escalante National Monument in what state?
Utah

14. Mesa Verde National Park is located in what landlocked state bordering New Mexico to the south?
Colorado

15. Carlsbad Caverns National Park is located in what state bordering Texas?
New Mexico

16. Guadelupe Mountains National Park is located in what state bordering Chihuahua and the Gulf of Mexico?
Texas

Round 8: Major Cities

1. What city in India is currently South Asia's most populous metropolitan area?

A Competitor's Compendium to the Geography Bee

Delhi

2. What city is the most populous metropolitan area in the world?
Tokyo

3. What city is the most populous in Nigeria, located in the southern region of the country?
Lagos

4. Shenzhen and Guangzhou are megacities in what country?
China

5. Jakarta, one of the most populous metropolitan areas in the world, is located on Java in what country?
Indonesia

6. What city is the most populous in Pakistan, in the country's south?
Karachi

7. What city in South America is more populous than Rio de Janeiro but less populous than Sao Paulo?
Buenos Aires

8. Sydney and Melbourne are major cities in what country?
Australia

9. Delhi, with an estimated population of nearly 33 million in 2025, is ahead of what other Indian city in population?
Mumbai

10. Dhaka is the largest metropolitan area and urban agglomeration in what country bordering the Bay of Bengal?
Bangladesh

11. Auckland is the largest city in what country whose capital is Wellington?
New Zealand

12. Pretoria and Cape Town are major cities in what country?
South Africa

13. Dubai is the most populous metropolitan area in what country bordering the Persian Gulf?
United Arab Emirates

14. What city is the most populous metropolitan area in North America, with a population of about 20 million people?
Mexico City

15. Chennai and Hyderabad are major metropolitan areas in the southern region of what country?
India

16. Delhi, which has grown from 4.4 million in 1975 to 24 million in 2015, is the most populous city in what South Asian country?
India

A Competitor's Compendium to the Geography Bee

National Competition Preliminaries

1. The Camaguey Archipelago belongs to what country in the Greater Antilles?
   Cuba

2. The Fortress of Suomenlinna is located in what Nordic country?
   Finland

3. Cairns and Brisbane are major cities in what state in Australia?
   Queensland

4. The Qadisha Valley is located in what country bordering the Mediterranean Sea to the west?
   Lebanon

5. Taman Negara National Park is located in what country?
   Malaysia

6. Docampado Bay borders what country to the east with the cities of Cali and Cartagena?
   Colombia

7. The Alhambra is a fortress in Granada in what country?
   Spain

8. Kota and Korba are cities in the northern region of what country?
India

9. Lundey Island is located in what country bordering the Norwegian Sea?
Iceland

10. The Bernese Alps and Lepontine Alps are located in what country bordering Lake Constance?
Switzerland

11. Rila National Park and Stara Zagora can be found in what country?
Bulgaria

12. Kosti is a major city in what country containing the Sanganeb Lighthouse?
Sudan

13. Kasakna National Park is west of the Muchinga Mountains in what country?
Zambia

14. The Louisiade Archipelago belongs to what country bordering the Ysabel Channel?
Papua New Guinea

15. Segou, Kayes, and Mopti are major cities in what country?
Mali

16. Songni Mountain National Park is located in the central region of what country?
South Korea

17. The Rabnabad Islands and Donmanick Islands are located in what country bordering the Bay of Bengal?
Bangladesh

18. Villeurbanne and Grenoble are cities in what country bordering the Mediterranean Sea?
France

19. Trinidad Island is located in what country where Mogotes Point and Cape Tres Puntas can be found?
Argentina

20. The Hall Peninsula and Incognita Peninsula border Frobisher Bay in what country?
Canada

21. San Lorenzo and Ciudad del Este are major cities in what landlocked country?
Bolivia

22. The Gulf of Taganrog is part of what inlet of the Black Sea?
Sea of Azov

23. The Irbe Strait separates Saaremaa from what country?
Latvia

24. Bonito is a peak in what country bordering Guatemala?

Honduras

25. Cueva de las Maravillas, meaning "Cave of Wonders" is a site in what country where Beata Island can be found?
Dominican Republic

26. The Chukchi language is a Paleosiberian language spoken by the Chukchi people in what country?
Russia

27. The djembe is a type of goblet drum that originated from what region in Africa?
West Africa

28. The dinar is the currency of what country whose prime minister is Jaber Mubarak al-Sabah?
Kuwait

29. Malalcahuello-Nalcas National Reserve is located in what country?
Chile

30. Jigme Dorji National Park is located in what country bordering India and China?
Bhutan

A Competitor's Compendium to the Geography Bee

2013 National Geographic Bee Finals Competition

1. What country is bordered by Burkina Faso and Libya?
Niger

2. What country is bordered by Thailand and Cambodia?
Laos

3. What country is bordered by Turkey and Saudi Arabia?
Iraq

4. What country is bordered by Panama and Nicaragua?
Costa Rica

5. What country is bordered by Somalia and Kenya?
Ethiopia

6. What country is bordered by North Korea and Myanmar?
China

7. What country is bordered by Angola and Malawi?
Zambia

8. What country is bordered by Mexico and Guatemala?

Belize

9. What country is bordered by Belarus and Romania?
Ukraine

10. What country is bordered by Ecuador and Peru?
Colombia

11. What country is bordered by Lithuania and Germany?
Poland

12. Mountaineer Gerlinde Kaltenbrunner became the first woman to climb the world's 14 highest peaks without using supplemental oxygen. The final peak in her expedition was K2, located on the border between China and what other country?
Pakistan

13. K2 lies in what mountain range that is an extension of the Hindu Kush mountain system?
Karakoram Range

14. Geneticist Spencer Wells explores the human past by collecting DNA samples from around the world. One group studied is a segment of the Bushman population that lives in the desert region west of the Caprivi Strip in what country?
Namibia

15. Northern Namibia is the site of one of the world's largest wildlife parks, which is centered on a large salt pan. Name this salt pan.

Etosha Pan

16. Botanist Joseph Rock, posing here in the 1920's with the king of Muli, made many expeditions to the upper Salween River region near the Tanggula Range. This range lies in what present day country?
China

17. Joseph Rock studied plant life throughout southwest China, including what province that borders Myanmar, Laos, and Vietnam?
Yunnan

18. Conservationists Beverly and Derek Joubert are raising awareness about the decline of big cats in the wild. These explorers are studying lions in the Okavango Delta in what country?
Botswana

19. The Okavango Delta feeds into what lake to the south?
Lake Ngami

20. Adventurer Kira Salak, pictured here with her guide, walked in the footsteps of 19th century explorer Hugh Clapperton through the historic Tripolitania region of what country?
Libya

21. Name the historic region south of Tripolitania that includes the Saharan oases of Sabha and Marzuq.
Fezzan

22. Biologist Roman Dial climbs a mountain eucalyptus tree while performing an ecological survey in the Hume Plateau, located in the Great Dividing Range in what county?
Australia

23. Part of Dial's research took place near the Australian Alps, which stretch from New South Wales into what other state?
Victoria

24. In 1938, archaeologist Matthew Stirling uncovered giant stone heads from the Olmec people in the village of La Venta, located on the Isthmus of Tehuantepec in what country?
Mexico

25. These stone heads were moved to museums in the capital of the state of Tabasco. Name this city.
Villahermosa

26. Conservationist Mike Fay displays his field journal while documenting one of the world's largest remaining concentrations of elephants in Zakouma National Park. This park is located southeast of the city of N'Djamena in what country?
Chad

27. The city of N'Djamena is located at the confluence of the Logone River and what other river?
Chari River

28. Climber Bradford Washburn mapped and photographed mountains throughout the world, including Mount Logan, the highest peak in what Western Hemisphere country?
Canada

29. Situated in Yukon's southwest corner, Mount Logan is located in what subrange of the Coast Ranges?
St. Elias Mountains

30. In the late 1800's, Norwegian explorer Fridtjof Nansen led an expedition to the North Pole, reaching a record northern latitude. On his retreat, Nansen crossed Franz Josef Land, an archipelago that belongs to what present day country?
Russia

31. Nansen was rescued from Franz Josef Land and returned to the city of Vardo in northeastern Norway. What sea lies between Franz Josef Land and Vardo?
Barents Sea

32. Conservation photographer Steve Winter has documented the tension between wildlife and humans in Gunung Leuser National Park, located northwest of the Barisan Mountains in what island country?
Indonesia

33. The Barisan Mountains are located on what island?
Sumatra

34. Boa constrictors are found throughout tropical Central and South America, and are now an invasive species on the westernmost island of the Lesser Antilles. Name this island.
Aruba

35. Diamonds are exported from what city in South Africa?
Port Elizabeth

36. Port Elizabeth is located on what bay?
Algoa Bay

37. Sugar is exported from what city in Indonesia?
Surabaya

38. Surabaya is located at the western end of what strait?
Madura Strait

39. What port city is a busy cruise ship terminal in the United Kingdom?
Southampton

40. Southampton is located north of what island in the English Channel?
Isle of Wight

41. Coffee is exported from what city in Ecuador?
Guayaquil

42. What island lies opposite of the mouth of the Guayas River at the head of the Gulf of Guayaquil?
Puna Island

43. Automobiles are exported from what city in South Korea?
Busan

44. Busan lies at the mouth of South Korea's longest river. Name this river.
Nakdong River

45. Glazed tiles are exported from what city in Spain?
Valencia

46. Valencia is located near the mouth of what river in Spain?
Turia River

47. Refined petroleum is exported from what city in Egypt?
Alexandria

48. Alexandria is northeast of a basin that is located in the Libya Desert. Name this basin.
Qattara Depression

49. Electronics are exported from what city in China?
Qingdao

50. Qingdao lies on the southern coast of what peninsula?
Shandong Peninsula

51. Cacao is exported from what city in Brazil?
Salvador

52. Salvador is located on what bay in Brazil?

Todos os Santos Bay

53. Natural gas is shipped from what port city in Australia?
Darwin

54. Darwin lies just south of the Tiwi Islands. Name the largest of these islands.
Melville Island

55. Textiles are exported from what port city in Iran?
Bandar 'Abbas

56. Bandar Abbas lies north of the largest island in the Strait of Hormuz. Name this island.
Qeshm

57. The whale shark can be found seasonally in the Mesoamerican reef system near Isla Holbox off the coast of what peninsula?
Yucatan Peninsula

58. The city of Porto lies on what river that flows west into the Atlantic Ocean?
Douro River

59. The Charles Darwin Research Station is located near Puerto Ayora on what island that is the second largest in the Galapagos?
Santa Cruz

60. Machu Picchu overlooks what river that flows into the Ucayali River?

A Competitor's Compendium to the Geography Bee

Urubamba River

61. Costa Rica's Corcovado National Park is located on what peninsula that borders Dulce Gulf?
Osa Peninsula

62. The Lemaire Channel lies northeast of what sea that borders Ellsworth Land and Alexander Island?
Bellingshausen Sea

63. The Pearl Islands, with a storied past involving Spanish conquistadors and buccaneers, are located in what large gulf?
Gulf of Panama

64. Shackleton is buried on what island that is administered by the United Kingdom and lies at the northeast corner of the Scotia Sea?
South Georgia

65. Gray whales migrate to Magdalena Bay, located off the coast of what Mexican state that includes the city of La Paz?
Baja California Sur

66. Name the capital of the British Overseas Territory that is located off the coast of Argentina?
Stanley

67. The oldest form of Tai Chi traces its roots back to a village near the city of Zhengzhou, located near the Yellow River in what province?

Henan

68. According to legend, a Taoist priest named Zhang Sanfeng is credited as the creator of Tai Chi. Some believe that he developed many of the movements of Tai Chi at a monastery in the Wudang Mountains, located just south of what river that is a tributary of the Yangtze?
Han River

69. Place these countries in order according to their land area, from largest to smallest: Iran, Yemen, Egypt.
Iran, Egypt, Yemen

70. Which of those countries - Iran, Egypt, Yemen - has the highest population density?
Egypt

71. Place these major cities in order according to their longitude, from west to east: Libreville, Lagos, Bangui.
Lagos, Libreville, Bangui

72. Place these lakes in order according to their surface area from largest to smallest: Nicaragua, Kyoga, Balkhash.
Balkhash, Nicaragua, Kyoga

73. Which of those lakes – Nicaragua, Kyoga, Balkhash - lies closest to the Equator?
Lake Kyoga

74. Place these major cities in order according to their rainfall, from most to least: Dublin, Tokyo, Sofia.
Tokyo, Dublin, Sofia

75. Which of those cities – Dublin, Tokyo, Sofia - has a marine west coast climate?
Dublin

76. Place these major rivers in order according to their length, from longest to shortest: Magdalena, Indus, Yellow.
Yellow, Indus, Magdalena

77. Which of those rivers – Magdalena, Indus, Yellow - has its mouth located farthest south?
Magdalena River

78. Place these major cities in order according to their latitude, from north to south: Budapest, Prague, Zagreb.
Prague, Budapest, Zagreb

79. Which of those cities - Budapest, Prague, Zagreb - has the largest population?
Budapest

80. Place these countries in order according to their GDP per capita: Argentina, Denmark, Slovenia.
Denmark, Slovenia, Argentina

81. Which of those countries has a constitutional monarchy as its form of government?
Denmark

82. Place these islands in order according to their land area, from largest to smallest: Cyprus, Halmahera, Taiwan.

Taiwan, Halmahera, Cyprus

83. Which of those islands - Cyprus, Halmahera, Taiwan - is crossed by the Tropic of Cancer?
Taiwan

84. Place these countries in order according to their population density, from the most densely populated to the least: Belgium, Slovakia, Italy.
Belgium, Italy, Slovakia

85. Which of those countries has the highest percent urban population?
Belgium

86. African penguins live in the waters around southern Africa, including what island that lies at the mouth of Table Bay?
Robben Island

87. Another colony of African penguins can be found on the mainland near Simon's Town, a seaside resort located on what bay?
False Bay

88. Which city is the odd one out? Alicante, Coimbra, Malaga, Valladoid.
Coimbra - the other cities are in Spain, while Coimbra is in Portugal

89. Which sect is the odd one out? Shaktism, Shia, Sufism, Sunni.

Shaktism - the other branches are branches of Islam, while Shaktism is a branch of Hinduism

90. Which city is the odd one out? – Bengkulu, Bandung, Medan, Banda Aceh.
    Bandung - the other cities are located on Sumatra, while **Bandung is located on Java**

91. Which city is the odd one out? – Barranquilla, Buenaventura, Cartagena, Cienaga.
    Buenaventura - the other cities are located on the Caribbean Sea, while Buenaventura is located on the Pacific Ocean

92. Name the Odd Item Out: Altay Mountains, Rhodope Mountains, Kopetdag Mountains, Tian Shan.
    Rhodope Mountains - the other mountain ranges are located in Asia, while the Rhodope Mountains are located in Europe

93. Name the Odd Item Out: Lake Van, Lake Tuz, Lake Urmia, Lake Iznik.
    Lake Urmia - the other lakes are located in Turkey, while **Lake Urmia is located in Iran**

94. Name the Odd Item Out: Sabarmati River, Krishna River, Narmada River, Tapi River.
    Krishna River - the other rivers flow into the Arabian Sea, while the Krishna River flows into the Bay of Bengal

95. What river is a heavily trafficked waterway, was once controlled by the Ancient Gauls, and rises in the Langres Plateau to then flow about 480 miles?
Seine River

96. The Pont de Normandie, one of the longest cable-stayed bridges in the world, spans the mouth of what river?
Seine River

97. The Cathedral of Notre Dame is located on an island in what river rising in the Langres Plateau?
Seine River

98. Located in the Simien Mountains, Ras Dejen is the highest peak in what country?
Ethiopia

99. One of the world's largest deposits of rare earth elements, which are used in the production of many high tech gadgets, is located near the largest city in China's Inner Mongolia Autonomous Region. Name this city.
Baotou

100. A capital city on the Arabian Peninsula located at about 7,200 feet receives its water supply from an aquifer that is forecast to run dry in the next decade. Name this capital city.
Sanaa

101. Name the oil-rich exclave that lies just north of the mouth of the Congo River.
Cabinda

102. Because Earth bulges at the Equator, the point that is farthest from Earth's center is the summit of a peak in Ecuador. Name this peak.
    Chimborazo

A Competitor's Compendium to the Geography Bee

2015 National Geographic Bee Finals Competition

1. Cadillac Ranch, which features ten Cadillacs covered in spray paint graffiti, is located near the largest city in the Texas panhandle. Name this city.
   Amarillo

2. This dinosaur sculpture attracts tourists to South Dakota's second largest city. Name this city.
   Rapid City

3. A 31-foot-tall statue of Paul Bunyan is located in Maine in the largest city on the Penobscot River. Name this city.
   Bangor

4. The world's largest rubber stamp is located in Ohio in the largest city on the Cuyahoga River. Name this city.
   Cleveland

5. Eight-foot-tall painted cowboy boots can be found throughout the largest city in Wyoming. Name this city.
   Cheyenne

6. Cupid's Span, a 60-foot-tall bow and arrow sculpture, is located near Fisherman's Wharf and the Presidio in one of California's largest cities. Name this city.
San Francisco

7. A 76-foot-tall statue of an oil driller stands at Expo Square in Oklahoma's second largest city. Name this city.
Tulsa

8. A 40-foot-tall blue bear peers through the windows of a convention center in the largest city in the Front Range. Name this city.
Denver

9. An 800-pound statue of a dog stands guard at a trailer company in the largest port city in Georgia. Name this city.
Savannah

10. This giant spoon and cherry decorate a sculpture garden in Minnesota's largest city. Name this city.
Minneapolis

11. Wynton Marsalis performed at a festival in a city on Narragansett Bay, located on the southern end of Aquidneck Island. Name this city.
Newport

12. Name the Indonesian mountain range that stretches the length of the island of Sumatra where Titan arum can be found.
Barisan Mountains

A Competitor's Compendium to the Geography Bee

13. Name the country where two European scientists invented flashlight powder in 1887 near the city of Potsdam.
Germany

14. Name the river along which the Havasupai tribe lives on the Coconino Plateau.
Colorado River

15. Plastic marine debris is threatening wildlife on what small group of U.S. – administered islands south of Kure Atoll?
Midway Islands

16. What U.S. city is located at the mouth of the Patapsco River?
Baltimore

17. Name the highest mountain in Oregon.
Mount Hood

18. Louisiana's flag depicts what animal?
Pelican

19. Name the largest city in the San Joaquin Valley.
Fresno

20. Wilmington, North Carolina, is located on what river?
Cape Fear River

21. What crop is the most widely planted field crop in the United States?
Corn

22. What U.S. city is located at the head of Cook Inlet?
Anchorage

23. The Pecos River is a tributary of what other river?
Rio Grande

24. What is the term for a narrow strip of land that connects two larger landmasses?
Isthmus

25. What Iowa City is located across the Missouri River from Omaha?
Council Bluffs

26. The lowest point in Idaho is on what river?
Snake River

27. The Mesabi Range contains a large deposit of what metal-bearing mineral?
Taconite

28. What city is located near the confluence of the McKenzie and Willamette Rivers?
Eugene

29. The Keweenaw Peninsula juts into what large lake?
Lake Superior

30. How many U.S. states have populations over 15 million?
    Four

31. Name the largest city on the Big Island of Hawaii.
    Hilo

32. What mountain is the highest point in Maine?
    Mount Katahdin

33. What is the term for a line on a map that connects points of equal temperature?
    Isotherm

34. What river connects Lake St. Clair to Lake Erie?
    Detroit River

35. The Seward Peninsula borders what strait?
    Bering Strait

36. What legume is the official state crop of Georgia?
    Peanut

37. What U.S. city is located where the Fox River enters Lake Winnebago?
    Oshkosh

38. What lake is the lowest point in Vermont?
    Lake Champlain

39. The Powder River Basin is best known for its large deposits of what fossil fuel?
    Coal

40. Name the largest city on Humboldt Bay.
    Eureka

41. The Niobrara River is a tributary of what larger river?
    Missouri River

42. How many U.S. states border the Gulf of Mexico?
    Five

43. What major city is located near the mouth of the Genesee River?
    Rochester

44. What dam created Lake Mead?
    Hoover Dam

45. South Carolina's flag depicts what tree?
    Palmetto Tree

46. Australia's Geographe Bay, a popular vacation spot, is located about 150 miles south of which state capital city?
    Perth

47. Italy's chief port city is on a gulf of the Ligurian Sea. Name this port city.
    Genoa

48. A desert, which has a name meaning "Black Sands," covers more than 70 percent of Turkmenistan. Name this desert.
    Karakum Desert

49. Fishing and sheep raising are the chief economic activities in a group of Danish islands that lies northwest of the Shetland Islands. Name these islands.
Faroe Islands

50. Lake Gatun, an artificial lake that constitutes part of the Panama Canal system, was created by damming which river?
Chagres

51. Which city on the Lena River is an important fur-trading center in Siberia?
Yakutsk

52. The returned sarcophagus will be housed in the Grand Egyptian Museum on the outskirts of what city located 3 miles southwest of Cairo?
Giza

53. The painting 'A Sunday Afternoon on the Island of La Grande Jatte' depicts a scene on an island located in what river?
Seine River

54. French fries may have originated along the Meuse River near the city of Liege in which country?
Belgium

55. What is the most densely populated country in Central America?
El Salvador

56. What is the capital of the Marshall Islands?
   Majuro

57. What is the official language of St. Lucia?
   English

58. Lake Ohrid straddles the border between Albania and what neighboring country?
   Macedonia

59. Peru's chief seaport is just west of Lima. Name this city.
   Callao

60. What is the official religion of Bangladesh?
   Islam

61. The atoll of Funafuti is part of which country?
   Tuvalu

62. What river flows through the city of Astrakhan?
   Volga River

63. What is the official currency of Liechtenstein?
   Swiss franc

64. Tripolitania is a historic region in which country?
   Libya

65. Name the second largest city on the island of Tasmania.
   Launceston

66. What is the official religion of Sri Lanka?

Buddhism

67. The island of Gozo is part of which Mediterranean country?
Malta

68. What is the capital of Timor-Leste?
Dili

69. What is the currency of Mongolia?
Mongolian togrog

70. A causeway connects Saudi Arabia and what country?
Bahrain

71. Name the only national capital city on the island of New Guinea.
Port Moresby

72. What is the official language of Togo?
French

73. Mariupol, a city located at the mouth of the Kalmius River, is located on what sea that is an arm of the Black Sea?
Sea of Azov

74. Helsingor's strategic located on a narrow strait allowed Danish kings to collect tolls from passing ships. Name this strait.
Oresund

75. A Russian island that straddles 180 degree longitude is one of the most biodiverse in the Arctic and is the world's northernmost UNESCO World Heritage Site. Name this island.
Wrangel Island

76. In 2014, the government of India established a new state out of the northwestern part of Andhra Pradesh. Name this new state.
Telangana

77. The Strait of Canso separates mainland Canada from what island?
Cape Breton Island

78. What Central Asian capital city is located northwest of the densely populated Fergana Valley?
Tashkent

79. If completed, the proposed Grand Inga Dam would become the world's largest hydropower plant. This dam would be built near Inga Falls on which African River?
Congo River

A Competitor's Compendium to the Geography Bee

USA Geography Olympiad/iGeo Resources

This chapter is designed to help you prepare for the USA Geography Olympiad and the International Geography Olympiad.

What you'll need to know:

World Geography

U.S. Geography

Physical Geography

Cultural Geography

Economic Geography

Historical Geography

A Competitor's Compendium to the Geography Bee

Current Events

Political Geography

Topography and Elevations

Interpreting Maps

Using Geographic Diagrams

What resources you'll need with you:

- Atlas (National Geographic Atlases are the best)
- World Maps (Again, national geographic – go to http://education.nationalgeographic.com/education/mapping/outline-map/?ar_a=1)
- U.S. Maps (You can find this in link provided and NG Atlases
- State/Province/Administrative Division Maps (United States, Canada, India, China, Australia, etc. – you can find these in link provided and in NG Atlases)
- Online blank maps of continents, countries, and the world
- Current Event websites with geo-related information

A Competitor's Compendium to the Geography Bee

- Earth Science textbooks (Borrow them from your library or school)

There are more than 1000 questions here to help you prepare. These are not included as part of the questions found in this book.

Study hard!

Links:

http://www.geographyolympiad.com/regionals/qualifying-exams-answer-keys/

http://www.geographyolympiad.com/nationals/sample-questions-geography-challenge-nationals/

http://www.geoolympiad.org/fass/geoolympiad/previous.shtml

A Competitor's Compendium to the Geography Bee

Geo Statistics

Here are some geographical statistics to help you gain some more knowledge. You don't have to memorize the exact populations of countries, or which comes after which in terms of area unless it is the top 10 countries.

These are to help you get an idea of the basics of these countries, and to provide insight into major languages, religions, and states/province in major countries.

Here, I have also included geographical statistics for India, for any NSF Geography Bee participants looking for more information.

Largest Country in the World: Russia

Smallest Country in the World: Vatican City

Most Populous Country in the World: China

Least Populous Country in the World: Vatican City

Largest Continent in the World: Asia

Smallest Continent in the World: Australia/Oceania

Most Populous Continent in the World: Asia

A Competitor's Compendium to the Geography Bee

Least Populous Continent in the World: Australia/Oceania

Largest Island by Area: Greenland

Highest Point in the World: Mount Everest

Lowest Point in the World: Mariana Trench (Challenger Deep)

Lowest Surface Point in Asia: Dead Sea

Lowest Surface Point in Africa: Lake Assal

Lowest Surface Point in South America: Laguna del Carbon

Lowest Surface Point in North America: Death Valley

Lowest Surface Point in Europe: Caspian Sea

Lowest Surface Point in Australia/Oceania: Lake Eyre

Lowest Surface Point in Antarctica: Byrd Glacier

Highest Point in Asia: Mount Everest

Highest Point in South America: Cerro Aconcagua

Highest Point in North America: Denali

A Competitor's Compendium to the Geography Bee

Highest Point in Africa: Kilimanjaro

Highest Point in Europe: El'brus

Highest Point in Antarctica: Vinson Massif

Highest Point in Australia/Oceania: Mount Kosciuszko

Tallest Mountain Above and Below Sea Level: Mauna Kea

Highest Mountain Above Sea Level: Mount Everest

Longest Mountain Range Above Sea Level: Andes Mountains

Longest Mountain Range Above and Below Sea Level: Mid-Ocean Ridge

Largest Cave Chamber: Sarawak Chamber, Gunung Mulu National Park, Malaysia

Longest Cave System: Mammoth Cave

Lowest Point in the Pacific Ocean: Mariana Trench (Challenger Deep)

Lowest Point in the Atlantic Ocean: Puerto Rico Trench

A Competitor's Compendium to the Geography Bee

Lowest Point in the Indian Ocean: Java Trench

Lowest Point in the Arctic Ocean: Molloy Deep

Largest Ocean: Pacific Ocean

Smallest Ocean: Arctic Ocean

Largest Sea in the World: Coral Sea

Longest River in the World: Nile River

Longest River in Africa: Nile River

Longest River in South America: Amazon River

Longest River in Asia: Yangtze River

Longest River in North America: Mississippi-Missouri River

Longest River in Europe: Volga River

Largest River Drainage Basin in the World: Amazon River

Largest River Drainage Basin in South America: Amazon River

Largest River Drainage Basin in Africa: Congo River

A Competitor's Compendium to the Geography Bee

Largest River Drainage Basin in North America: Mississippi-Missouri River

Largest River Drainage Basin in Asia: Ob-Irtysh River

Largest Lake in the World: Caspian Sea

World Population: 7,290,000,000 (7.29 billion)

Most Densely Populated Country in the World: Monaco

Least Densely Populated Country in the World: Mongolia

Countries Sharing the Greatest Number of Borders with other Countries: China and Russia

Tallest Building in the World: Burj Khalifa, Dubai, United Arab Emirates

Tallest Pyramid in the World: Great Pyramid of Khufu, Egypt

Longest Wall in the World: Great Wall of China

Longest Road in the World: Pan-American Highway

A Competitor's Compendium to the Geography Bee

Longest Railroad in the World: Trans-Siberian Railroad, Russia

Longest Bridge in the World: Danyang-Kunshan Grand Bridge, China

Longest Suspension Bridge in the World: Akashi-Kaikyo Bridge, Japan

Tallest Road Bridge in the World: Millau Viaduct, France

Largest Reservoir by Surface Area: Lake Volta, Volta River, Ghana

Largest Reservoir by Volume: Lake Kariba, Zambia/Zimbabwe

Tallest Dam in the World: Nurek Dam, Vakhsh River, Tajikistan

Largest Hydroelectric Power Station in the World: Three Gorges Dam, China

A Competitor's Compendium to the Geography Bee

Longest Submarine Cable in the World: Sea-Me-We 3 (Southeast Asia – Middle East – Western Europe) cable, connects 33 countries on four continents

Hottest Place in the World: Dalol, Danakil Depression, Ethiopia

Coldest Place in the World: Ridge A, Antarctica

Hottest Recorded Air Temperature: Furnace Creek Ranch, Death Valley, California

Coldest Recorded Air Temperature: Antarctica

Wettest Place in the World: Mawsynram, Meghalaya, India

Driest Place in the World: Arica, Atacama Desert, Chile

Largest Hot Desert in the World: Sahara Desert, Africa

Largest Cold (Ice) Desert in the World: Antarctica

Largest Canyon in the World: Grand Canyon, Colorado River, Arizona

Largest Coral Reef Ecosystem in the World: Great Barrier Reef, Australian Pacific Coast

A Competitor's Compendium to the Geography Bee

Greatest Tidal Range in the World: Bay of Fundy, Canadian Atlantic Coast

Tallest Waterfall in the World: Angel Falls, Venezuela

Deepest Lake in the World: Lake Baikal, Russia

Oldest Lake in the World: Lake Baikal, Russia

Strongest Recorded Wind Gust: Barrow Island, Australia

Northernmost Permanently Inhabited City/Settlement in the World: Alert, Canada

Northernmost Permanently Inhabited City/Settlement with more than 1,000 People: Longyearbyen, Svalbard (Norway)

Largest City North of the Arctic Circle: Murmansk, Russia

Southernmost City in the World: Ushuaia, Tierra del Fuego (Argentina)

Southernmost Town in the World: Puerto Williams, Chile

A Competitor's Compendium to the Geography Bee

Northernmost Capital City in the World: Reykjavik, Iceland/Nuuk, Greenland

Northernmost Point of Land on Earth: Kaffeklubben Island, Greenland

Northernmost Active Volcano on Earth: Beerenberg, Jan Mayen (Norway)

Northernmost Lake in the World: Lake North Pole, North Pole

Northernmost Crater Lake in the World: Lake Lapparjarvi, Finland

Southernmost Navigable Body of Water: Bay of Whales, Ross Sea (Antarctica)

Westernmost Point on Land in the World: Attu Island, United States

Easternmost Point on Land in the World: Caroline Island, Kiribati

Point Farthest from the Earth's Center: Chimborazo, Ecuador

Highest Navigable Lake in the World: Lake Titicaca, Bolivia and Peru

Most Remote Capital City in the World: Wellington, New Zealand and Canberra, Australia (this may change since Noumea is the capital of New Caledonia, a French overseas territory scheduled to vote for independence from France between 2014 and 2018)

Most Remote City with a Population Greater Than 500,000 People: Honolulu, Hawaii (United States)

Most Remote Airport in the World: Mataveri International Airport, Easter Island (Chile)

A Competitor's Compendium to the Geography Bee

About the Author

Keshav Ramesh is a 13-year old author of 18 books and a geography enthusiast. Keshav participated in the 2015 and 2016 Connecticut State Geographic Bees.

In addition to geography, he participated in the Scripps National Spelling Bee when he was in fourth and fifth grade. Keshav was a 2016 MathCounts State Finalist and competed in the 2014/2015 AMC 8 and 2016 AMC 10.

Keshav plays piano and tennis, and his interests include math, geography, tennis, basketball, music, writing, and reading.

You can follow him on Twitter @keshavramesh1

Visit his websites, www.prepgeobee.blogspot.com and www.geobeeworld.blogspot.com for geography bee tips, information, questions, and how to prepare for the Bee.

A Competitor's Compendium to the Geography Bee

Bibliography

You should use these resources to help you in your preparation for the National Geographic Bee. Use the websites listed here well!

National Geographic Atlas of the World, Tenth Edition. N.p.: National Geographic Society, 2014. Print.

Infoplease. Infoplease, n.d. Web. 2015. <http://www.infoplease.com/>.

National Geographic Kids Ultimate Adventure Atlas of Earth. N.p.: National Geographic Society, 2015. Print.

National Geographic Kids Ultimate Globetrotting World Atlas. N.p.: National Geographic Society, 2014. Print.

National Geographic GeoBee. National Geographic, n.d. Web. 2015. <www.nationalgeographic.com/geobee>.

National Geographic Kids Almanac 2016. N.p.: National Geographic Society, 2015. Print.

National Geographic Kids World Atlas. N.p.: National Geographic Society, 2013. Print.

National Geographic Kids United States Atlas. Washington, D.C.: National Geographic Society, 2012. Print.

Wikipedia. Wikimedia Foundation, n.d. Web. 2015. <https://www.wikipedia.org/>.

Wojtanik, Andrew. *The National Geographic Bee Ultimate Fact Book: Countries A to Z*. Washington, D.C.: National Geographic Society, 2012. Print.

"World Geography." *FactMonster World*. FactMonster, n.d. Web. 2015. <http://www.factmonster.com/world.html>.

World Atlas. World Atlas, n.d. Web. 2015. <http://www.worldatlas.com/>.

National Geographic Kids. United States Encyclopedia: America's People, Places, and Events. Washington, D.C.: National Geographic Kids, 2015. Print.

www.ingramcontent.com/pod-product-compliance
Lightning Source LLC
Chambersburg PA
CBHW050153230526
45470CB00001B/76